Facilitating Sustainable Innovation through Collaboration

Facilitating Sustainable Innovation through Collaboration

Joseph Sarkis · James J. Cordeiro ·
Diego Vazquez Brust
Editors

A Multi-Stakeholder Perspective

 Springer

Editors
Joseph Sarkis
Clark University
950 Main Street
Worcester, MA 01610-1477
USA
jsarkis@clarku.edu

James J. Cordeiro
The College at Brockport
State University of New York (SUNY)
350 New Campus Drive
Brockport, NY 14420-2914
USA
jcordeir@brockport.edu

Diego Vazquez Brust
The ESRC Centre for Business
    Relationships, Accountability,
    Sustainability and Society (BRASS)
Cardiff University
55, Park Place
Cardiff
Wales CF10 3AT
United Kingdom
vazquezd@cardiff.ac.uk

ISBN 978-90-481-3158-7   e-ISBN 978-90-481-3159-4
DOI 10.1007/978-90-481-3159-4
Springer Dordrecht Heidelberg London New York

Library of Congress Control Number: 2010922294

Cover image: Cover image taken and reworked by Gary Oldknow (http://www.deepvisual.com/). The
cave wall painting symbolizes "layers of meaning. . .tradition was innovation".

Printed on acid-free paper

Springer is part of Springer Science+Business Media (www.springer.com)

Contents

Contributors

Marlen Arnold TUM Business School, Technische Universität München, Alte Akademie 14, Freising, 85350, Germany, marlen.arnold@wi.tum.de, http://www.food.wi.tum.de; http://www.nanu-projekt.de

Mikael Backman International Institute for Industrial Environmental Economics, IIIEE, Lund University, Lund, Sweden, mikael.backman@iiiee.lu.se

Hans Th. A. Bressers Centre for Clean Technology and Environmental Policy (CSTM), University of Twente, Enschede, 7500 AE, The Netherlands, j.t.a.bressers@utwente.nl

Kjell-Erik Bugge Saxion University of AppliedSciences, P.O. Box 501, Deventer, 7400 AM, The Netherlands, k.e.bugge@saxion.nl

Robyn Bushell Centre for Cultural Research, University of Western Sydney, Locked Bag 1797, Penrith South DC, NSW 1797, Australia, r.bushell@uws.edu.au

Per Christensen Department of Development & Planning, Aalborg University, Fibigerstraede 13, Aalborg E, 9220, Denmark, pc@plan.aau.dk

James J. Cordeiro Department of Business Administration and Economics, The College at Brockport (SUNY), 350 New Campus Drive, Brockport, NY 14420, USA, jcordeir@brockport.edu

Augusto Cuginotti Master's Programme "Strategic Leadership towards Sustainability", Blekinge Institute of Technology, Karlskrona, Sweden, acuginotti@gmail.com

Nicole Darnall Department of Environmental Science & Policy, George Mason University, MSN 5F2, Fairfax, VA 22030, USA, ndarnall@gmu.edu

Theo de Bruijn Center for Clean Technology and Environmental Policy (CSTM), School of Management and Governance, University of Twente, Enschede, 7500 AE, The Netherlands, Theo.deBruijn@utwente.nl

AnnaKarin Djupenström Business Excellence Specialist, Group Business Excellence Team, Stora Enso Oyj, Helsinki, Finland, annakarin.djupenstrom@storaenso.com

María Laura Franco-García Centre for Clean Technology and Environmental Policy (CSTM), University of Twente, Enschede, 7500 AE, The Netherlands, m.l.francogarcia@utwente.nl

Ian Hill European and Regional Development Consultancy, Cockermouth, UK, ian.nweurope@btinternet.com

Peter S. Hofman Nottingham University Business School China, Ningbo 315100, China, Peter.Hofman@nottingham.edu.cn

Björn Johnson Department of Business Studies, Aalborg University, Fibigerstraede 4, Aalborg E, 9220, Denmark, bj@business.aau.dk

Linda M. Kamp Delft University of Technology, Jaffalaan 5, Delft 2628 BX, The Netherlands, l.m.kamp@tudelft.nl

Reine Karlsson TEM Foundation at Lund University, Klostergatan 12, Lund, 222 22, Sweden, reine.karlsson@tem.lu.se

Martin Lehmann Department of Development and Planning, Aalborg University, Fibigerstraede 13, Aalborg E, 9220, Denmark, martinl@plan.aau.dk

Haiying Lin University of Waterloo, Faculty of Environment, Business and Environment, EV1-231, 200 University Avenue West, Waterloo, Ontario, N2L 3G1 Canada, h45lin@uwaterloo.ca

Karen Marie Miller Strategic Sustainable Development, Blekinge Institute of Technology, Karlskrona 371 79, Sweden, karenmariemiller@yahoo.ca

Frank O'Connor Ecodesign Centre, Cardiff Business Technology Centre, Senghennydd Road, Cardiff, CF244AY, UK, frank@edcw.org

Bill O'Gorman Centre for Enterprise Development and Regional Economy (CEDRE), Waterford Institute of Technology, Carriganore Campus, Waterford, Ireland, wogorman@wit.ie

Simon O'Rafferty Ecodesign Centre, Cardiff Business Technology Centre, Cardiff, 24 4AY, UK, simon@edcw.org

Joseph Sarkis Graduate School of Management, Clark University, Worcester, MA 01610-1477, USA, jsarkis@clarku.edu

Jennifer Scott School of Natural Sciences, University of Western Sydney, Locked Bag 1797, Penrith South DC NSW, Australia, jenny.scott@uws.edu.au

Bruce Simmons School of Natural Sciences, University of Western Sydney, Locked Bag 1797 Penrith South DC NSW, Australia, b.simmons@uws.edu.au

Deborah M. Steketee Department of Sustainable Business, Aquinas College, Grand Rapids, MI 49506, USA, stekedeb@aquinas.edu

Freek van der Pluijm Master's Programme "Strategic Leadership towards Sustainability", Blekinge Institute of Technology, Karlskrona, Sweden, freek@strategicsustainabledevelopment.eu

Diego Alfonso Vazquez Brust ESRC Centre for Business Relationships Accountability Sustainability and Society (BRASS), Cardiff University, 55 Park Place, Cardiff, CF10 3AT, UK, VazquezD@cardiff.ac.uk

Jan Venselaar Research Group Sustainable Business Operation, Avans University of Applied Sciences, PO Box 1097, Tilburg</Cty, 5004, The Netherlands, j.venselaar@avans.nl

Friederike Welter RUREG, Jönköping International Business School (JIBS), PO Box 1026, Jönköping, 551 11, Sweden, Friederike.welter@ihh.hj.se

List of Figures

List of Tables

List of Boxes

Chapter 1
Facilitating Sustainable Innovation through Collaboration

Joseph Sarkis, James J. Cordeiro, and Diego Alfonso Vazquez Brust

Abstract Innovation, sustainability, and collaboration are all related in their efforts to manage multiple dimensions of organizational and institutional policies and practices. This chapter provides an overview of the three topics and their relative importance to overall advancement of sustainability through innovations. Collaboration is necessary to achieve this goal and various collaborative arrangements and stakeholders in these arrangements are discussed. The chapter also introduces and discusses the various remaining chapters in this book and presents summaries, insights and linkages amongst these chapters.

Keywords Innovation · Sustainability · Collaboration · Stakeholders · Triple helix

1.1 Defining Sustainability, Defining Innovation

Considerations of facilitating sustainable innovation through collaboration must start by recognizing that sustainability and innovation are flexible terms and that actors can interpret them in different ways, leading to possible misunderstanding, conflict or misappropriation of terms for vested interests (Hajer, 1995). Since collaboration requires the use of dialogue and reasoned argumentation to foster mutual understanding and this in turns requires trust and transparency in the use of terms (Habermas, 1996). With this in mind, we provide some relevant definitions and contrasts below.

The terms "sustainability" and "innovation" are widely used by social coalitions promoting development. However not everyone committed to sustainable development strives for it in the same way (Seyfang & Smith, 2007). Apparent consensus on

J. Sarkis (✉)
Graduate School of Management, Clark University, Worcester, MA, 01610-1477, USA
e-mail: jsarkis@clarku.edu

J. Sarkis (eds.), *Facilitating Sustainable Innovation through Collaboration*,
DOI 10.1007/978-90-481-3159-4_1, © Springer Science+Business Media B.V. 2010

the promotion of sustainability and innovation often masks significant differences on the meaning and implications that actors assign to these words (Hajer, 1995).

Sustainability is often associated with the Brundtland Report definition for Sustainable Development "following the needs of the present without compromising the ability of future generations to meet their own needs" (WCED, 1987). Yet sustainable development is not the same as sustainability. According to Dresner (2002) sustainability represents the goal of a "sustainable society" a term used for the first time in 1974 by the World Council of Churches to define a society where environmental, economic and social concerns are integrated. Sustainability is thus an ideal equilibrium condition, while sustainable development is a pathway from today's unsustainable socio-technical systems toward such equilibrium.

It is necessary also to distinguish between weak and strong sustainability. Weak sustainability sees all forms of capital as substitutes for one another (for example, substitution of technology for natural capital) whereas strong sustainability reflects the stance that natural materials and services cannot be duplicated or replaced by man-made capital.

While sustainability is about equilibrium and permanence, innovation is about changing the way things are done. It is a form of learning to solve specific problems in a highly differentiated and volatile context (Dicken, 2006), and implies uncertainty about effects. Innovation aimed at providing new technologies as a solution to protect ecosystems can shift the use of resources and impact social and natural systems in new and unexpected ways.

The scale of innovation is also relevant. *Incremental innovations* are small scale, progressive refinements of existing products, processes or ideas that occur relatively continuously and often unnoticed. On a larger scale, *radical product-process innovations* are unpredictable events that drastically change existing products and processes. When there is a "cluster" of such innovation, a widespread shift in the socio-economic system can be triggered. *Systems innovation* is radical innovation involving changes in technology that create entirely new sectors of the economy through a combination of radical and incremental technological accompanied by organizational innovation (Dicken, 2006). Finally, *paradigm-shifting* or *disruptive innovation* involves large scale, pervasive changes in the techno-economic paradigm involving the mode of management and production (i.e., the introduction of electric power (Loorbach & Rotmans, 2006)).

Are any of these types of innovation more conducive to sustainability? Perspectives differ; one view is that incremental innovation in product and business practices, when accumulated over time can steer economic activity onto a sustainable pathway. Adherents of this view also believe that this type of innovation can be created through learning by doing and its consequences predicted to some extent through forecasting and risk assessment. The socio-technical perspective, in contrast, argues that the predictability of incremental innovation is its great weakness, since it permits development of political and economic processes resisting necessary changes in production and consumption, regulation and infrastructure (Hoogma, Kemp, Schot, & Truffer, 2002), and therefore that only disruptive, unpredictable innovation can change our unsustainable development patterns (Berkhout,

2002; Geels, 2005). These are still issues that need addressing and research guidance in these areas will be valuable from an institutional investment perspective.

Finally, we must recognize differences on the preferred path to sustainability. *Narrow sustainability* follows a technological path focused on improving the environmental efficiency of production through ongoing innovation and environmental management. *Broad sustainability* on the other hand, is based on the view that technological innovation aimed at providing mere fixes to environmental and social problems is not sustainable because artificially prolongs intrinsically unsustainable socio-economic structures (changing to sustain the status quo). Real sustainable innovation is seen as involving a change in socio-economic structures and in our relationship with the natural environment (Dryzek, 2005). This perspective views the world as a socio-biological system with resource and pollution flows between poor and affluent regions of countries. The path to sustainability is seen as involving paradigm-shifting innovation leading to the minimization of inequities and injustices through changes to existing political and economic systems.

1.2 Innovation for Sustainable Enterprise

Many levels and dimensions of innovation need to be considered in the sustainability context. These range from abstract and relatively intangible innovations related to institutional and policy development to more tangible innovations related to durable product and technological innovation.

Incorporating innovation into a model of sustainable development is notably difficult (Newman, 2005). Innovation for sustainability, whether it is incremental, radical, narrow or broad, will be complex and multidimensional; a single organization is unlikely to have the resources to effectively innovate in this arena. The process is socially and institutionally embedded with multiple actors – each of whom may have a different perspective and interest – and which can occur at expanding levels of scale, each with deeper, larger and more unpredictable consequences in the equilibrium of ecological and socioeconomic systems (Rihani, 2002). Therefore, more interdisciplinary research and collaborative efforts among organizations, their partners and their stakeholders is needed to better understand the effects of innovation and create a more effective innovation environment.

Organization theorists writing on innovation have long recognized this need for collaborative integration as the following quote demonstrates:

> Innovation is not the enterprise of a single entrepreneur. Instead, it is a network-building effort that centers on the creation, adoption, and sustained implementation of a set of ideas among people who, through transactions, become sufficiently committed to these ideas to transform them into 'good currency'.
>
> (Ven De Ven, 1986, p. 601).

Within the technology and operations management fields too, it is recognized that for radical innovation to occur, as in the case of sustainability innovation, a focus on collaboration is required (Clark & Wheelwright, 1993). Collaborative efforts within the corporate (for-profit) sector may occur within organizations, groups and teams that span functional and organizational boundaries such as supply chain partnerships and strategic alliances with other corporate entities.

Sustainability issues also range beyond the strategic or operational concerns for innovation by for-profit organizations. Beyond the corporation and its partners (e.g., green supply chains (Sarkis, 2006)), various external stakeholders need to be incorporated usefully into the innovation enterprise, including government agencies, universities, non-governmental organizations, and even communities. Such a multi-stakeholder group perspective is usefully represented by the acronym MAGPI (for Market, Academia, Government, Public, and Industry). In this context, research and evaluation can occur from a variety of perspectives, and collaboration can take various other forms – formal negotiations, voluntary agreements, stakeholder dialogues, networking, green supply chains, multiple-partner projects, information-sharing – and can vary extensively in terms of their size, membership, goals and actions (Fadeeva, 2005). Poncelet (2001) reviews some of the benefits of such multi-stakeholder collaborations, including more efficient resource utilization, speedier, more participative, and more creative solutions.

Collaborations for innovation can be reviewed from the perspective of various stakeholders:

The Perspective of For-Profit Corporations. This perspective has been most central to the innovation literature and management organization theory and is the most mature research stream within the still developing field of sustainability innovation. Much of the innovation here may be considered entrepreneurial in spirit with the ultimate goal of economic sustainability of the firms and their supply chains. The triple bottom line is important in this context.

While corporations are driven by profit, other organizations and stakeholders may be driven by non-economic factors depending on their stakeholder constituency. We consider three salient ones next.

The Perspective of Public Policy Makers. Policy-related practice and research focuses on institutional aspects of the public-private linkage, incorporating a variety of perspectives related to budgeting, planning, formation, execution and auditing of the public programs that support this type of linkage. The degree to which a program's designs are put into practice and the objectives achieved provide key performance evaluation points. Public policy makers also determine the budgeting and controls for helping guide sustainable innovation.

The Perspective of a Public R&D Body or Laboratory or University. The role of these organizations in the sustainable development enterprise will range from basic to applied research development as these organizations typically have the necessary scientists and research resources to drive innovation. They are the repositories of old and new knowledge. Exploitative (applied) and exploratory (basic) research are two of the primary resources that they offer to collaboration.

The Perspective of the Non-governmental Organization (NGO). Innovation is arguably a key capability of successful NGOs (Fyvie & Ager, 1999) which tend to have a deep insight into community needs and local factors. Such attributes can be better harnessed through partnerships with firms and governments creating "new social compacts" (Brugmann & Prahalad, 2007) to deliver socially and environmentally responsible products/services and intervention that increases community empowerment and self-reliance.

Ongoing theory development is vital for sustainable innovation, as is the development of supporting methodologies, frameworks and tools that integrate stakeholder inputs into the collaborative innovation process. In this respect, corporate social responsibility, sociological and policy frameworks are valuable. Beyond standard diffusion of innovation and knowledge based collaborative efforts, stakeholder, institutional, and ecological modernization theories may also explain collaborative efforts for sustainability innovation. Institutional innovation in sustainability also includes learning (van der Kerkhof & Wieczorek, 2005), adaptive management (Foxon, Reed, & Stringer, 2009), and prevention (Johnson, Hays, Center, & Daley, 2004) that is managed by the variety of stakeholders.

Our book seeks to touch on these many points of collaboration, sustainability and innovation. We now introduce the book contents to help introduce the reader to the various topics covered and provide an integrative perspective of the book.

1.3 Introduction to the Content of the Book

Since this is a multi-disciplinary topic with a variety of levels and dimensions to be examined, the book's content and organization could have been presented in numerous ways. Table 1.1 provides the various concepts and issues examined by the authors that helped guide our categorization and organization. Chapters are arranged in the order in which they appear in the book. Since we are focusing on a multi-stakeholder perspective, we have also defined the major stakeholders involved in each chapter. Another noteworthy characteristic is the use the multiple levels of analysis, notably geographic region and organization/institutional characteristics (e.g., supply chains). Overall, the scope of coverage ranges from broad and inclusive modes of collaboration (e.g., regional systems and triple helix – collaboration of authorities, industry and universities) to those that are relatively specific (e.g., specific organizations).

Karlsson et al.'s Chapter 2 reflects on the contribution – and limitations – of triple helix approaches along with insights on the role of empowerment, open dialogue and investment thinking to foster intelligent innovation. The chapter builds on the experience of the authors in four action-research case studies on sustainability oriented collaboration between regional authorities, universities and businesses in the Oresund region of Sweden and Denmark. In the analysis of the cases the authors use a variety of metaphors to highlight the relations between factors influencing the success of sustainable innovation processes. The entrepreneur delivering intelligent innovation, for instance, is presented as a "driver" with a need for appropriate instruments, a steering wheel and an inspiring vision. The chapter concludes that transformative learning and entrepreneurs' engagement in radical renewal activities

Table 1.1 Overview and content of chapters

Chapter title	Stakeholder focus	Scope	Methodology	Theoretical perspective
Sustainable Considerations and Triple Helix Cooperation in Regional Innovation Systems	Industry, academia and government	Regional	Case studies	Triple helix Investment thinking
Partnerships and Sustainable Regional Innovation Systems: Special Roles for Universities?	Academia and industry	Regional	Conceptual	Expanding partnerships to triple helix
Obstacles to and Facilitators of the Implementation of Small Urban wind Turbines in the Netherlands	Industry, government, academia	Regional and local	Conceptual	Socio-technical theory
Regional Sustainability, Innovation and Welfare Through an Adaptive Process Model	Government	Regional	Empirical	Systems theory Adaptation Triple helix
FOCISS for an Effective Sustainable Innovation Strategy	Industry, government and academia	Organizational and supply chain	Case studies	Innovation prioritization
Emergence of Sustainable Innovation: Key Factors and Regional Support Structures	Government and industry	Regional	Conceptual	Systems theory Multi-level innovation
Disruption or Sustenance? An Institutional Analysis of the Sustainable Business Network in West Michigan	Government and industry	Local	Empirical	Institutional analysis
Regional Perspectives on Capacity Building for Ecodesign – Insights from Wales	Industry and public	Regional and supply chain	Case studies	Systems theory (System failure)
Fostering Responsible Tourism Business Practices Through Collaborative Capacity Building	Government and industry	Local and regional	Case studies	Engagement and capacity building
Design and Decision-Making: Backcasting Using Principles for Cradle-to-Cradle	Industry	Supply chain	Conceptual	Cradle-to-Cradle Life cycle innovation
Corporate Strategies for Sustainable Innovation	Industry	Supply chain	Case studies	Strategic change and sustainability
A New Typology of Strategic Alliances and Its Strategic Implications	Industry and public (NGOs)	Global/national	Conceptual	Institutional theory Resource based view
Towards Sustainability by Negotiated Agreements Between Industrial Sectors and Government – Mexican Case	Industry and government	Regional/national	Empirical	Voluntary mechanisms; negotiated agreements

are both required to move toward successful sustainable innovation, and suggests that open-minded triple-helix collaboration can facilitate the mobilization of sufficient concerted "investments" in radical, innovative sustainable development "experiments".

The role of universities within various sustainability and innovation networks is still in its infancy. Using two Danish sustainability innovation network case studies, Lehmann et al. compare and contrast the role of universities in Chapter 3. They find that university partnership and collaboration is dependent on a variety of "capital" factors and issues. They also see the role as contingent upon various political and institutional factors that help in its development. An important finding is that these collaborative networks are evolving. One network, born as a triple-helix type arrangement, is defined as Public-Private-Academic partnerships (PPAP). The other described the academic role in the helix as a very low level functioning position and defined it as a Public-Private partnership (P3) for this reason. Part of the explanation of these differing roles may be based on the strategic positioning of the university within these two network studies. One had the university leading the network; the other had the university playing a very peripheral role. Specific operational functions and roles also provide some pertinent insights. The roles of the university in these sustainability innovation partnerships range from the knowledge leader and basic research to technology transfer and dissemination. The authors describe the cases and the results of their analysis within their Greening Triangle collaborative framework which is useful in further understanding the roles of the academic stakeholders in these collaborative networks as well as other stakeholders.

Socio-technical theory and systems are powerful explanatory tools to help understand how collaboration at various levels and by various stakeholders can serve as barriers or enablers to success of sustainable innovations. Using these theories Kamp, in her Chapter 4 shows how coordination among multiple levels of the socio-technical system is required for success. Missing elements at any level can doom the development or diffusion of these technologies. Using the case of urban wind technology she goes through a multi-functional framework based on socio-technical systems theory, and using the observations from these functions, Kamp identifies the barriers and enablers for this specific innovative technology. Also, using the semantics of system dynamics modeling of vicious and virtuous cycles she qualitatively describes the various elements that could prove useful for a quantitative system dynamics tool. While qualitative, this chapter provides significant insight into how collaboration for sustainable innovation can be quantitatively modeled.

Bugge et al.'s Chapter 5 introduces a decision-making policy tool: The Adaptive Model CRIPREDE. This tool – designed and applied in six countries as part of a major collaborative action-research project funded by the European Union – facilitates and stimulates collaborative interaction leading to regional learning, innovation, and transformation of networks contributing to a more sustainable future. As in Chapters 2 and 7, the theoretical foundations for this tool draw on triple helix and entrepreneurship theories (in this case entrepreneurial ecosystems theory). These concepts can be effectively combined with Triple-P principles as in Lehmann et al.'s Chapter 3. The authors first analyze the development of the Adaptive Model and its theoretical basis, then focus on the results and policy implications of its highly interactive process of application in six very different (in terms of political, cultural, economic features) regions across the EU: City Triangle in The Netherlands; Cumbria in UK, Latgale in Latvia, Novo Mesto in Slovenia, Siegen-Wittgenstein in Germany, and South East Ireland. The chapter emphasizes the overriding importance of engaging and empowering regional stakeholders – "the drivers and owners of the regional developmental process" – in particular the triple helix networks of industry, authorities, and universities. The authors conclude that finding the right involvement of relevant stakeholders is a key challenge for the implementation of partnership-based collaboration in sustainable regional development. They suggest that collaboration needs to be pragmatic and at the same time have a strategic long-term perspective and framework. Stakeholders' involvement should accordingly be tailor-made to match expectations and possibilities and based on clearly identified responsibilities, added value, and synergy.

Policy and decision making tools for sustainability, especially with a broad variety of stakeholders are very uncommon. Venselaar introduces a valuable decision tool, called FOCISS, for small and medium sized enterprises in Chapter 6. This tool reflects the fact that while companies need to learn the various complexities of sustainability to effectively develop, implement and improve upon sustainable innovations, small and medium sized companies typically do not have the necessary capacities and knowledge to do this. A tool like FOCISS helps them to evaluate and prioritize factors that will help companies focus in on the most sustainable solution for internal innovations. The collaborative appeal of this tool is that it is general enough to integrate various internal cross-functional stakeholders as well as external inter-organizational stakeholders into the planning and decision making process. The tool is meant to simplify the complexities, but certain critical steps are to be followed for success. Lessons learned from the application of this tool to a number of case study companies provide insight into various pitfalls and lessons.

Hoffman and De Bruijn (Chapter 7) use Regional Innovation systems theory to understand how innovations evolve and what key explanatory factors contribute to the emergence and diffusion of innovations in 10 cases of Dutch firms with observed sustainable innovation processes to analyze. As is the case in Chapters 2 and 8, the authors emphasize the importance of radical, disruptive innovation for sustainable development. However, although the nature of innovation processes is addressed, the focus of the chapter is on the analysis of the relations between firms and the regional support structures that facilitated innovation processes, in particular on the study of gaps between the needs identified within firms' innovation processes and the type of functions provided by support structures. The authors conclude that – especially for SMEs – demand articulation remains a major barrier as users are often only involved when the innovation is ready to enter the market, while regional support functions in this respect are deficient. Moreover, SMEs have major difficulty interpreting and anticipating sustainability policies and regulations at local and national levels, leading to innovations that face major regulatory barriers or are unable to cope with policy changes. The chapter proposes that some functions in the current support structure could more effectively build regional support systems.

These functions include stimulation of demand articulation and vision development, supply of strategic intelligence that SMEs cannot obtain in house, and provision of interfaces between policy and business that allow firms to better cope with policy uncertainty and to anticipate policy developments.

In Chapter 8, Steketee focuses on the analysis of types of innovation network structures and institutional arrangements contributing toward sustainability. The author uses Institutional Analysis Design (IAD) to study the contribution of "game changing" disruptive innovation and "market-deepening" sustaining innovation to the sustainable development of West Michigan (USA), a region which appears to be successfully avoiding the downward spiral of other regions in the Great Lakes by replacing its "rust belt" with a "green" belt. The author attributes such success to the existence of a West Michigan "network of networks" dedicated to sustainable business. This network of networks is open to learning and transfer of knowledge but does not follow the conventional model of leadership, where a vision has been developed, a strategy designed and a team assembled to implement the strategy. On the contrary this is a system of overlapping networks and leaders that emerged to respond to the increasing pace of change and the disruptive circumstances which present themselves as a result of that change. The case study further supports two related hypotheses. First, business-led disruptive innovation, rather than sustaining innovation, is more effective in fostering transformative, sustainable regional development. Second, institutional arrangements supporting collaboration through networks are central to regional success in this transformation as individual firms' competency in innovation leverages sustainability as an organizing logic for regional development.

In Chapter 9, O'Rafferty and O'Connor, also use systems' theory, in this case combined with a capacity building framework, to identify enablers and blockers for the diffusion of a specific technology, analyzing four case studies of SMES that participated on a recent regional eco-design initiative in Wales (Great Britain). The cases are used as a means to explore strategies for public intervention to foster sustainable innovation and regional development in Wales through eco-design. The chapter addresses themes underrepresented in both the innovation literature and sustainable development literature such as the role of design as an innovative process, a collaborative process and a business strategy, and – in particular – eco-design's capability to support sustainable regional development. The chapter stands out from others in this book in terms of its focus on incremental innovation. In contrast to Steketee or Hoffman and De Bruijn, O'Rafferty and O'Connor argue that the cumulative impact of incremental innovations on long term economic development and social change can be equal or greater than radical innovations. Regarding theoretical foundations, the authors support public intervention as a response to a regional systems' failure to deliver sustainable innovation. The chapter shows similarities with others in this book when it identifies some of the determinants of successful collaborations: a new dialogue on the structure and content of public interventions (Karlsson et al., Chapter 3), a history of previous collaboration or trust (Franco-Garcia and Bressers, Chapter 14), open-minded collaboration and reciprocal learning (Karlsson et al., Chapter 3), regional authorities support in developing an infrastructure for linkages and co-operation between actors and agents (Hoffman and De Bruijn, Chapter 7; Steketee, Chapter 8) and the use of flexible and evolving intervention models (Bugge et al., Chapter 5).

Simmons et al. (Chapter 10) report on government-stakeholder collaboration by focusing on two collaborative research and development projects in Australia. Both projects were funded by the New South Wales State Government and built on partnerships that were established between university researchers at the University of Western Sydney and local government bodies (similar to case studies in Chapter 3). Specifically, the authors report on two cases of sustainability learning programs tested in the private the public sectors based on a tiered system of engagement, process, and performance. The programs – the Gumnut Awards Environmental Management Program for the Caravan and Camping Industry Association of New South Wales and the Sea Change for Sustainable Tourism program in Manly Beach, a residential Sydney suburb – targeted sustainability advances using social and environmental tools, and exemplified the potential for effective social change processes resulting from government collaboration with other relevant stakeholders. Using surveys and supporting thematic data analyses, the authors provide a rich account of the benchmarks, successes, barriers and limitations of each project and review the processes and partnership efforts involved, including, importantly, future challenges perceived by respondents, and methods utilized for removing barriers and resistance to participation in the programs.

The study has some noteworthy aspects. As Hoffman and De Bruijn, Chapter 7, noted the focus on SMEs is welcome, as SMEs are an important focal constituency. They typically lack incentives, information and resources (especially time and personnel) to go beyond compliance with mandated environmental regulation and thus stakeholder collaboration initiatives to enhance their sustainability efforts are especially warranted. The Australian setting of the cases is also interesting from the sustainability viewpoint given the country's low population density. Most importantly, the resulting rates of environmental management system (EMS) adoption by the SMEs in the two programs were much higher (18–50%) than the Australian national rates of EMS adoption (6–7%). These results provide both reasons for optimism as well as many valuable insights that should help similar stakeholder collaborations in the future.

Chapter 11 by van der Pluijm et al. provides an insightful look into a novel technique for implementing the cradle-to-cradle concept which is gaining momentum in Western Europe and has established footholds in Japan and the U.S., and particularly in the Netherlands. The cradle-to-cradle concept seeks to learn from nature and to design using principles that emphasize the conversion of waste into food, the use of solar energy inputs, and the celebration of diversity. As such, it facilitates organizational transition toward enabling a societal infrastructure by participating in cyclical supply chains – a valuable complement to the green supply chain approach to organizational collaboration for sustainability.

The specific contribution of van der Pluijm, Miller and Cuginotti to the cradle-to-cradle literature focuses on the integration of cradle-to-cradle design within a systems approach that permits analysis from a strategic sustainable development perspective. After usefully comparing cradle-to-cradle design principles with FSSD (framework for strategic sustainable development) principles for sustainability, the authors integrate science-based principles with value-based principles as an asset to support backcasting using overarching sustainability constraints drawn from scientific principles for socio-economic sustainability. This framework is one that decision-makers can use flexibly to make mid-course corrections in the march toward a societal infrastructure that supports a targeted system in which all material flows are either part of a biological or a technical metabolism. This approach is noteworthy in that society is brought into the sustainable enterprise as a collaborative stakeholder alongside entrepreneurs and community builders, facilitating progress along the front of ecological modernization.

Arnold's study (Chapter 12) utilizes a case analysis of three companies to examine the implications of organizational, cultural, and external structural conditions for active corporate sustainability programs. The organizational factors include flexibility and active searching routines and knowledge transfer and the variety of corporate capabilities and patterns of action, the cultural factors include management's sustainable values, vision and norms and corporate entrepreneurship, while the external factors include market demands, competition, state regulation as well as stakeholder demands. These factors are used to predict the timing and intensity of strategic change from an environmental perspective within the organizations studied. The three companies studied are noteworthy for the different foci of their sustainability efforts. Novamont (Switzerland) focused on circular flows for recycling in the local economy, Bedminster (Italy), the recycling company focused on zero emission, while the Dutch company Phillips uses life cycle analysis integrated with the Eco Vision program.

Based on semi-structured personal and phone interviews (supplemented by written follow-up surveys) Arnold utilized content analysis to code and interpret data from multiple individual cases using the analytical framework of organizational, cultural and external factors described above and examined cross-company patterns. Importantly, the focus was not just on the present, but also on past decisions and patterns of action as well as future visions and planned actions. Her results provide valuable insights into corporate environmental strategic change processes and actions.

The focus of Lin and Darnall's Chapter 13, which is complemented by Arnold's study on corporate environmental programs, argues that notion of partnerships as a theme to sustainability is hardly new (see, for example, early work in the GIN context by Hartman, Hofman, and Stafford, 1999). Lin and Darnall take an important step toward redressing an important omission in the literature on alliances for sustainability. Integrating two theoretical perspectives from organization theory – institutional theory and the resource-based view of the firm – they assess firms' decisions to participate in strategic alliances that advance proactive corporate environmental strategies. Lin and Darnall examine and develop the logic of how the corporate quest for competency-building and legitimacy motivates the choice for corresponding competency-oriented or legitimacy-oriented alliances, which in turn lead to various environmental strategies on a spectrum ranging from reactive strategies (such as pollution control) to proactive strategies (such as the quest for clean technology). The authors develop two propositions for future research: competency-oriented alliances will tend to associate with more proactive environmental strategies while legitimacy-oriented alliances will tend to associate with less proactive environmental strategies. This chapter contributes to the strategic alliance literature by developing a useful framework to assess alliance formation in the environmental management context. To our knowledge, this is the first such work in this area and represents a welcome advance. In addition to providing corporate environmental researchers with two interesting and testable hypotheses, the chapter provides readers who are new to this area access to useful, state-of-the art reviews of the typology of corporate reactive and proactive environmental strategies (ranging from pollution control to pollution prevention to product stewardship to clean technology), and, discussion of the relevance of the institutional theory and resource-based perspectives for understanding these strategies.

Finally, Franco-Garcia and Bressers's Chapter 14 focuses on a developing country, one of the few in our book, and discusses the extent to which a policy-tool that had been developed for a specific European context can be transferred to the ostensibly different Latin-American environment. Specifically, Franco-Garcia and Bressers evaluate the extent to which Dutch experiences with negotiated agreements between firms and public authorities could be used as a tool to improve environmental policies and foster collaboration and innovation for sustainability in Mexico. The authors analyze the Mexican context both in terms of perceived effectiveness of environmental regulation/existing voluntary agreements and in terms of attitudes and opinions of key players in the Mexican industry regarding feasibility of negotiated agreements. Their findings show that there is good receptivity to the use of negotiated agreements both from the point of view of policy makers and industry leaders. The comparison with Dutch experiences shows no important gap between Mexican business leaders' expectations regarding results in terms of efficiency gains and positive side effects and the results obtained by negotiated agreements in the Netherlands. Mexico benefits from a history of trust and fair play between the industrial sector and the government; homogeneity or clear leadership in polluting industrial sectors. Polluting firms are also concerned with their public image and there is a widespread belief that the government will resort to other measures if negotiation fails. All the latter factors, which were determinant of success in The Netherlands, support the feasibility of using negotiated agreements as a collaborative strategy toward sustainability in Mexico.

1.4 Concluding Overview and Suggestions for Future Research Directions

Readers will appreciate the variety of methodological approaches taken by the authors. They include theoretical conceptualizations, practical and focused case studies, as well as broader empirical evaluations, and the theoretical or framework developments utilized derive from the organizational theory, policy studies, environmental research, research policy, and systems engineering disciplines. This characteristic further exemplifies the interdisciplinary research needed for both sustainability and innovation. The richness and diversity of approaches provides some critical and novel insights into collaboration for sustainability innovation that we hope will have an important place on the road map to global sustainable development.

While sustainable innovation is a key driver of sustainable enterprise, we must guard against being too sanguine about its potential. Fadeeva (2005), for example, shows that collaboration frequently falls short of expectations, and that the achievement of satisfactory results depends on a number of factors that might well be overlooked by collaborating partners. Future research needs to continue to explore these factors, and to focus especially on consensus-building techniques. Sustainable innovation research would also benefit from the incorporation of new techniques, such as, for example, that exemplified in recent work drawing on the insights of game theory (Lozano, 2007).

Future research would also benefit greatly from careful integration with the broader literature on management innovation. A useful and up to date review of this literature is provided by Birkinshaw, Hamel, and Mol (2008) who develop an innovation process framework built around successive stages of motivation, invention, implementation, theorization and labeling. The management innovation literature, while focused principally on for-profit organizations, also has important implications for non-profits and other stakeholders in the sustainable development enterprise, especially in a post corporate world (Limerick, Cunnington, & Crowther, 2002). In addition to the management theoretical perspectives represented in this volume, recent contributions from relatively new areas from organizational economics in a sustainability context such as corporate governance themes derived from agency theory might prove to be quite valuable, for example recent work on the incentives provided to top managers for advancing sustainable goals (Cordeiro & Sarkis, 2008).

To achieve a better understanding of alternative pathways to sustainability and innovation dynamics, sustainable innovation research might also benefit from exploring interdisciplinary theoretical frameworks drawing on new approaches to sustainability and innovation developed in the social and political sciences, in particular transition theory literature (Geels, 2005). [1] An example is the "Pathways to Sustainability" framework (Leach et al., 2007) which builds on theories of complexity, deliberative democracy, resilience, path-dependency, non-linear dynamics

---

[1]Transition theory literature emphasizes the interdependency of institutions and infrastructures defining societal systems and sub-systems, thus creating different types of lock-in that stymie innovation (path dependencies for technological and social developments such as, existing competencies, past investment, habits, regulation, social norms, dominant discourses). In particular, Transition theory argues that the stability and cohesion of societal systems is created and maintained through institutional *regimes* (sets of practices, rules, norms and shared assumptions that focus on system *optimization* rather than system *innovation* (Geels, 2005; Loorbach & Rootmans, 2006)

and uncertainty[2] with a view to understanding the links between ecological sustainability and technology and poverty reduction and social justice. This framework sees the world as a complex array of constantly changing interactions between ecological, social, political and technological processes and actors. While collaboration and innovation are the heart of the pathways to sustainability, it is not always the case that collaboration leads to success. In order to foster sustainable innovation, collaboration has to be reflexive and dynamic, a constant process of identification of, and adaptation to change through renegotiation of solidarities and interdependences (Leach et al., 2007; Scoones et al., 2007).

A significant limitation of this volume is its almost exclusive focus on the developed world. Research on innovation in grassroots movements highlights that the five billion people living in developing countries are salient sources and beneficiaries of innovation (Pathak 2008; Seyfang & Smith 2007), challenging the dominant top-down, North-South approach for innovation dissemination. There are valuable opportunities for reverse knowledge transfer of innovation generated in developing countries for example, that must be studied more closely. Initiatives such as the Honeybees network, originated in India and now spreading to 75 countries (Gupta, Sinha, & Koradia, 2003) or the Fab Labs initiative supported by MIT have resulted in highly successful local innovations in areas as diverse as bio-security and digital technology. Collaboration between universities, firms, authorities and grassroots movements from both developed and developing countries may be a pathway to innovative and sustainable solutions to local challenges while simultaneously redressing distributive injustices. Thus, collaboration on sustainable innovation should not only encompass regions within nations, but should also be sought with collaboration across nations. The development and transfer of this knowledge can occur in both directions and should be encouraged as such.

There is ample theoretical foundation for the development and evaluation of sustainability and innovation from the technology and innovation, institutional policy, and sustainable development literatures. Fruitful investigation of key innovation relationships and processes requires a multi-disciplinary and multi-stakeholder perspective. Despite the valuable contributions captured in this book's chapters, we still have a long way to go in this field.

Acknowledgments (The Role of the Greening of Industry Network) This book itself stands as an example of a multi-stakeholder collaborative project for the diffusion of innovations related to sustainability. The concept and raw material for this book evolved from a 2008 conference sponsored by the Greening of Industry Network (GIN) in the Netherlands.

Kurt Fischer and Johan Schot began the work of organizing the Greening of Industry Network in 1989, before its official launch at the first GIN conference in November 1991. Thus this book arrives at the 20th anniversary of the conceptualization of GIN. GIN is one of the oldest inter-disciplinary and cross-institutional (multi-stakeholder) organizations focusing on the greening and sustainability of organizations. GIN is a prime example of collaboration and innovation

---

[2]The Pathways to Sustainability framework sees sustainability as a property of non-equilibrium systems allowing the maintenance of basic systems survival functions: equity, wellbeing and environmental quality during dynamics transitions from one equilibrium state to another.

for sustainability. Its website (http://www.greeningofindustry.org/), defines the organization as "an international network of professionals from research, education, business, civil society organizations, and government, focusing on issues of industrial development, environment, and society, and dedicated to building a sustainable future." Its mission statement reads "The Greening of Industry Network develops knowledge and transforms practice to accelerate progress toward a sustainable society."

GIN is managed today by an international group of eight coordinators, including original members Kurt Fischer of The George Perkins Marsh Research Institute at Clark University, Theo de Bruijn of the Center for Clean Technology and Environmental Policy (CTSTM) at the University of Twente, and Somporn Kamolsiripichaiporn from Chulalongkorn University. Over the years GIN's conferences have been held on different continents to accommodate the hundreds of members of the network around the world. The GIN is a vital presence in the global sustainability discourse.

This book is comprised primarily of select papers from the GIN conference held on June 26–28 in Leeuwarden, The Netherlands. The main theme of the conference was "Facilitating Sustainable Innovations Sustainable Innovation as a Tool for Regional Development Innovation". Thus, a broad focus on sustainability, technology, sustainable development, and policy guides the contents of this book. Even though regional development was the topic, many levels of analysis were represented, and the conference provided an excellent opportunity for knowledge transfer, a critical element of collaborative innovation to occur. Only the best papers that fit within the topical objectives of the book were included.

Our thanks goes not only to the coordinators of GIN, but also to Springer Publishers for their confidence and support for this project. Special thanks go to Fritz Schmuhl and Takeesha Moerland-Torpey for their important role in helping us bring this project to successful completion. Finally, of course, without the fine work of the contributors, this book will not be possible. We hope readers will find the chapters useful and insightful.

References

Berkhout, F. (2002). Technological regimes, path dependency and the environment. *Global Environmental Change, 12,* 1–4.

Birkinshaw, J., Hamel, G., & Mol, M. J. (2008). Management innovation. *Academy of Management Review, 35*(4), 825–845.

Brugmann, J., & Prahalad, C. K. (2007). Co-creating business's new social compact. *Harvard Business Review, 85,* 80–90.

Clark, K. B., & Wheelwright S. C. (1993). *Managing new product and process development.* New York, NY: The Free Press.

Cordeiro, J. J., & Sarkis, J. (2008). Does explicit contracting effectively link CEO compensation to environmental performance? *Business Strategy and the Environment, 17*(5), 304–317.

Dicken, P. (2006). *Global shift: Reshaping the global economic map in the 21st century.* London: Routledge.

Dresner, S. (2002). *The principles of sustainability.* London: Earthscan.

Dryzek, J. (2005). *The politics of the earth.* Oxford: Oxford University Press.

Fadeeva, Z. (2005). Promise of sustainability collaboration – potential fulfilled? *Journal of Cleaner Production, 13*(2), 165–174.

Fyvie, C., & Ager, A. (1999). NGOs and innovation: Organizational characteristics and constraints in development assistance work in The Gambia. *World Development, 27,* 1383–1395.

Foxon, T. J., Reed, M. S., & Stringer, L. C. (2009). Governing long-term social-ecological change: What can the adaptive management and transition management approaches learn from each other? *Environmental Policy and Governance, 19*(1), 3–20.

Geels, F. W. (2005). *Technological transitions and system innovation: A coevolutionary and socio-technical analysis.* Cheltenham, UK: Edward Elgar.

Gupta, A. K., Sinha, R., & Koradia, D. (2003). Mobilizing grassroots' technological innovations and traditional knowledge, values and institutions: articulating social and ethical capital. *Futures, 35*, 975–987.

Habermas, J. (1996). *Between facts and norms: Contributions to a discourse theory of law and democracy.* Cambridge, MA: MIT Press.

Hajer, M. (1995). *The politics of environmental discourse: Modernization and the policy process.* Oxford: Oxford University Press.

Hartman, C. L., Hofman, P. S., & Stafford, E. R. (1999). Partnerships: A path to sustainability. *Business Strategy and the Environment, 8*(5), 255–266.

Hoogma R., Kemp, R., Schot, J., & Truffer, B. (2002). *Experimenting for sustainable transport: The approach of strategic niche management.* New York: Routledge.

Johnson, K., Hays, C., Center, H., & Daley, C. (2004). Building capacity and sustainable prevention innovations: A sustainability planning model. *Evaluation and Program Planning, 27*, 135–149.

Leach, M., Bloom, G., Ely, A., Nightingale, P., Scoones, I., Sha, E., et al. (2007). *Understanding Governance: Pathways to sustainability* (STEPS Working Paper 2). Brighton: STEPS Centre.

Limerick, D., Cunnington, B., & Crowther, F. (2002). *Managing the new organization: Collaboration and sustainability in the postcorporate world.* St Leonards, NSW: Allen and Unwin.

Loorbach, D., & Rotmans, J. (2006). Managing transitions for sustainable development. In X. Olshoorn & A. J. Wieczorek (Eds.), *Understanding industrial transformation: Views from different disciplines.* Dordrecht: Springer.

Lozano, R. (2007). Collaboration as a pathway for sustainability. *Sustainable Development, 15*(6), 370–381.

Newman, L. (2005). Uncertainty, innovation, and dynamic sustainable development. *Sustainability: Science, Practice, & Policy, 1*(2), 25–31.

Pathak, R. (2008). Grass-root creativity, innovation, entrepreneurialism and poverty reduction. *International Journal of Entrepreneurship and Innovation Management, 8*, 87–98.

Poncelet, E. C. (2001). "A kiss here and a kiss there": Conflict and collaboration in environmental partnerships. *Environmental Management, 27*(1), 13–25.

Rihani, S. (2002). *Complex Systems: Theory and Development Practice.* London: Zed Books.

Sarkis, J. (Ed.). (2006). *Greening the supply chain.* Berlin: Springer.

Scoones, I., Leach, M., Smith, A., Stagl, S., Stirling, A., & Thopson, J. (2007). *Dynamics systems and the challenge of sustainability* (STEPS Working Paper 1). Brighton: STEPS Centre.

Seyfang, G., & Smith, A. (2007). Grassroots innovations for sustainable development: Towards a new research and policy agenda. *Environmental Politics, 16*(4), 584–603.

van der Kerkhof, M., & Wieczorek, A. (2005). Learning and stakeholder participation in transition processes towards sustainability: Methodological considerations. *Technological Forecasting and Social Change, 72*, 733–747.

Ven De Ven, A. H. (1986). Central problems in the management of innovation. *Management Science, 32*, 590–607.

WCED. (1987). *Our common future, the world commission on environment and development.* New York: Oxford University Press.

Chapter 2
Sustainability Considerations and Triple-Helix Collaboration in Regional Innovation Systems

Reine Karlsson, Mikael Backman, and AnnaKarin Djupenström

Abstract Sustainability challenges imply that there are severe needs for intelligent innovation processes. This chapter presents four case studies on sustainability oriented collaboration; including experiences from advanced leadership training, the Øresund Science Region innovation system, mobility of sustainability expertise, as well as business developments for hardwood. Two cases employ "Triple-Helix" collaboration between companies, research and the public sectors. In a metaphor, the entrepreneur is presented as a "driver" with a need for appropriate instruments, a steering wheel and an inspiring vision. Investment and depreciation of capital are used in analogies that explain why investment thinking is relevant also within the environmental dimension. A sustainability oriented model of material recycling is used as a metaphor to clarify how the sustainability value of experience varies dependent on how and where it is used. Self-esteem, empowerment and freedom of action are found to be essential to facilitate transformative learning. In addition to open dialogue it is vital to mobilize sufficient concerted "investments" in "real life experiments" with new creative ideas. To build motivation to engage in renewal oriented innovation, it is important to elucidate the human sustainability advantages that are likely to evolve as a result of more knowledgeable innovation.

Keywords Triple helix · Investment · Recycling · Innovation · Metaphors

2.1 Introduction

Innovation and empowerment are major driving forces for sustainable development. In relation to this, an ambition of the European Commission is to enable regions to play more important roles, enhancing the value of undiscovered, insufficiently

R. Karlsson (✉)
TEM Foundation at Lund University, Klostergatan 12, Lund, 222 22, Sweden
e-mail: reine.karlsson@tem.lu.se

J. Sarkis (eds.), *Facilitating Sustainable Innovation through Collaboration*,
DOI 10.1007/978-90-481-3159-4_2, © Springer Science+Business Media B.V. 2010

understood or ineffectively used resources. Accordingly, regions are to become a breeding ground for innovation.

> In essence, sustainable development is a process of change in which the exploitation of resources, the direction of investments, the orientation of technical development and institutional change are all in harmony and enhance both current and future potential to meet human needs and aspirations.
>
>   (WCED, 1987, Chapter 2:1:15)

Sustainable development is a "journey", rather than a "destination". One innovation promoting perspective is found in how Vinnova, the Swedish Governmental Agency for Innovation Systems, promotes Triple-Helix business development processes. Vinnova aims to promote innovation and sustainable growth throughout Sweden.

> The term Triple Helix is used to describe the interaction between actors in the fields of business, science and politics which produces effective innovation systems.
>
>   (VINNOVA, 2002)

To enable significant renewal there is a need for radical innovation. Transformative developments and path-breaking open-minded dialogue are dependent on challenging personal and institutional development processes.

> It takes courage to start a conversation. But if we do not start talking to one another, nothing will change. Conversation is the way we discover how to transform the world, together
>
>   (Wheatley, 2002, p. 31)

One way to promote personal advancement is empowerment; see also (Lindström, 2003), e.g., through microcredits and freedom of action.

> Every single individual on earth has both the potential and the right to live a decent life. Across cultures and civilizations, Yunus and Grameen Bank have shown that even the poorest of the poor can work to bring about their own development.
>
>   (The Norwegian Nobel Committee, 2006)

There is a huge potential when the human capability is mobilized in sustainable innovation processes. This chapter presents four different sustainable innovation initiatives as a means to explore the themes outlined above and its relation to other factors leading to the emergence and success of intelligent innovation processes.

2.1.1 The Four Case Studies

1. Science Region – a collaborative initiative, by universities, businesses and public development actors in the Øresund Region, orientated toward revitalization.
2. Leadership Training – an internal activity within the forestry based company Stora Enso in collaboration with EFQM, the European Foundation for Quality Management.
3. Mobility of Experts – a Vinnova project aiming to promote contacts between national experts and local business actors, in two countryside regions.

4. Hardwood System – Supply Chain development, aiming to develop the Swedish hardwood business system in concert with promotion of sustainable forestry.

This chapter presents the four initiatives and the principal results from action research within these initiatives. The ambition is to summarize and compare the general patterns of learning from the four different processes, aiming to achieve conceptual understanding.

2.1.2 Background

Forestry and related industries represent a major Swedish business area. In the 1800s Sweden had significant deforestation and consequently a mandatory replanting following logging activities was introduced in a forestry law in 1903 – an early form of sustainable forestry. During the 1900s Sweden invested in development of large forestry related industries. (This is a background for Case 4.)

Sweden has a tradition in systems thinking, e.g., in development of electrical and telephone networks. One innovation basis is Ideon, in the old university city of Lund, which has become a regional center for advanced business development. It is a meeting place for visionaries, entrepreneurs and risk capital. It was founded in 1983, at a time of crisis in the region's main industries. Over the years more than 600 companies have advanced from vision to operative business. Today, Ideon houses about 250 companies with in total 3000 employees. This is one innovation base in the Øresund Science Region.

2.2 Theoretical Foundation

Vinnova suggests a Triple-Helix model to promote "sustainable growth", which may be interpreted as a revitalization of the original dualism of the *sustainable (versus) development* concept. The *sustainable growth* ambition has attracted a lot of attention. However, it is criticized from an environmental point of view, as being too close to acceptance of unrestrained business growth, see also Dryzek (1997) and Murphy (1994). To enable a long-term sustainable development there is a need for renewal oriented entrepreneurship. However, it is difficult to assess the real sustainability effect of initiatives that surpasses the boundaries of established experience.

To be successful an innovation system has to provide multifaceted support for entrepreneurship. The Triple-Helix concept aims to compile a varied support capability through involvement of actors from the private, public and research sectors, see Fig. 2.1. The innovation support is enhanced through parallel advancements in the three development spirals. The Triple-Helix aim is to enhance the business development process and the total result through constructive collaboration among the three spheres.

Fig. 2.1 Triple helix growth
process

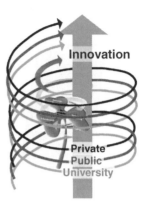

The main method to enhance the level of ability to sustain is to make "investments", i.e., present ventures to improve future conditions. There is a rich set of methods for economic assessment of investments. Analogous ways of thinking can also be applied in other dimensions.

- Money is invested to reduce future costs and to improve the production of added value.
- Reflection of social considerations can improve future social situations.
- Environmental load is often unavoidable in sustainability oriented investments.

One rationale is to invest present resources to improve the effectiveness of future resource production. However, nature does not allow an endless growth of material volumes; see also Murphy (1994). One key factor is to distinguish between qualitative and quantitative growth, and also between expenditure for consumption and money to investments.

2.2.1 The Basic Principle for Investments

In economics, the connection to the future is being accounted as investments and depreciation of capital, see Fig. 2.2. The environmental "accounting", e.g., LCA, focuses on the negative impact on future conditions; see also (Karlsson, 1994). The general environmental view does not observe the Fig. 2.2 cell that is marked "?". Consequently, resource improvements tend to be unseen from sustainability point of view.

In LCA, the output of a recyclable material is assessed as the avoided "load" in terms of alternative production activities that otherwise would have been needed. Formula 2.1 shows this principle in Fig. 2.3, depicting closed-loop recycling (Karlsson, 1998). The recovered resource value, R can be higher than the resource value N of the "natural" resource which is used as virgin raw material. The enhancement of the value level, from N to R, can be conceptualized as an investment of

Coverage with respect to time / Accounting	Effect during the time span under study	Change of situation from before to after	
		Deterioration	Improvement
Economic	Utility, profit	Production capacity changes	
		Depreciation	Investment
Environmental	Emissions, waste	Resource depletion	**?**

Fig. 2.2 Application of the economic investment perspective in the environmental dimension (Karlsson, 1994)

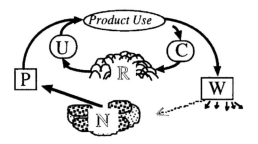

Activity systems
P – Primary production
U – Secondary upgrading
C – Collection
W – Waste disposal

Resource values
N – Raw material from nature
R – Recyclable material

Fig. 2.3   Basic flow diagram of a closed loop material market with recycling

present environmental load to reduce the future environmental load.

$$R = N + P - U \tag{2.1}$$

R = Sustainability resource value for recyclable material
N = Environmental resource value of raw material from nature
P = Environmental load from primary production
U = Environmental load from re-upgrading of recollected matter

The R-value is dependent on the future gain, not the historic "cost". In open loop recycling the sustainability value of the recovered material is dependent on the avoided load and resource consumption, in the receiving system; see Fig. 2.4 and Formula 2.2 (Karlsson, 1998).

$$R_{1,2} = N_2 + P_2 - U_2 \tag{2.2}$$

$R_{1,2}$ = Sustainability value for the output material that is recovered from System 1. All losses and handling up to the storage must be included in the assessment of System 1.

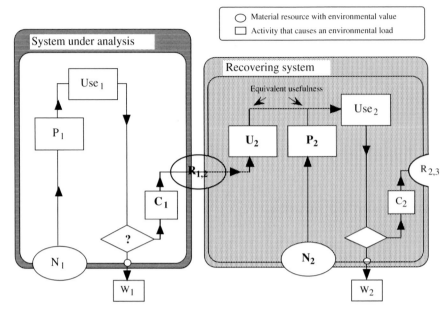

Fig. 2.4 Sustainability potential for product remains in open loop recycling and, specifically, the parameters in Formula 2

N_2 = Environmental value of the resource from nature which is used for the alternative production of an equivalent utility value, in System 2.

P_2 = Environmental load from the primary production evaluated as an alternative to the assessed recycling of the evaluated matter.

U_2 = Environmental load from the planned re-upgrading (in System 2) of the recovered matter, starting from the storage at the end-point of the System 1 assessment.

$R_{1,2}$ is dependent on the qualitative properties and the usefulness as a raw material in the receiving recycling based production, in comparison to a not recycling based alternative in System 2. If there is an abundance of alternative readily useful natural resources, then $R_{1,2}$ is low. If the alternative is to use rare resources and environmentally "costly" refinement, then $R_{1,2}$ is high. The upgrading from the quality of N_1 to a higher $R_{1,2}$ quality can be conceptualized as a sustainability investment.

Looking deeper, the above formulas only depict a part of the picture. It is common that there is quality degradation from the primary Use_1 material to the secondary Use_2 material (Karlsson, 1998). It is also difficult to know if the residual material really will be recycled to something truly useful. The ambition here is to illustrate the analogy that investment thinking is relevant also within the environmental dimension.

2.2.2 Analogies and Metaphors as Tools to Conceptualize New Aspects

The investment concept and the Fig. 2.4 model for recycling of material can be used as a conceptual analogy or metaphor also for nonmaterial dimensions, e.g., knowledge recycling.

Metaphors are enlightening as tools for transfer of understanding from an area which is known to a topic which is less familiar (Ortony, 1975). Metaphors such as Machine, Organism and Brain can be used to enhance the understanding of how an organization works (Morgan, 1986). The organizational theatre concept is used as a platform for analogically mediated inquiry and change (Meisser & Barry, 2007). Metaphors can be used to provide structure, to understand a process in a new light and to evoke emotions. However, metaphors that do not fit risk misrepresenting the information and confusing the scientific understanding (Carpenter, 2008).

2.2.3 The Entrepreneur as a "Driver"

An entrepreneur can be conceptualized as a "driver" of a development process. Figure 2.5 illustrates a driver's outlook with instruments and a steering wheel. The forest illustrates a development vision. Figure 2.5 represents a perspective analogous to the view found in the modern Swedish textbooks utilized for driver's education – in which many presented images depict the perspective as seen from a driver's point of view (from within the car). Such textbooks used to show an aerial view of cars on

People
can grow as "drivers",
it they are trusted and have freedom of action,
and appropriate
instruments for
"navigation".

Fig. 2.5 Possible working conditions for an entrepreneur; a driver with instruments for navigation and control, and a business development vision

a road (Karlsson & Luttropp, 2006). If we conceptualize entrepreneurs as "drivers" of their own projects and personal growth processes, then the experts' knowledge ought to be instruments on everybody's (own) dashboard.

When entrepreneurs are conceptualized as drivers with freedom of action it is obvious that there often are many roads and ways of driving that may lead to similar results.

2.3 Case Studies

The following presents the four initiatives and the results are summarized in Table 2.1. The authors can be conceptualized as developers of instruments for the various "drivers".

2.3.1 The Øresund Science Region

The Øresund Region has become a European hot spot for research, education, innovation and growth. In this cross-border region, the Danish-Swedish interaction presents opportunities for innovation. The Øresund Science Region (ØSR, 2008a) has developed the Øresund Model – a "Double Triple-Helix" model for growth, in a cross-border region. The ØSR Double Triple-Helix involves collaboration between two Triple-Helix processes, one in each county; see Fig. 2.6. ØSR bring together regional authorities, businesses and universities from the two different countries, with their different administrative and legal cultures, industrial landscapes and languages.

The Øresund Model aims to combine the forces of twelve universities in collaboration with the public sector and numerous companies. Some of the features are:

- border-crossing cooperation between a large number of actors;
- a double Triple-Helix system uniting the main regional actors;
- including NGOs and an entrepreneurial academy into the concept;
- the size of it: 12 universities – six science parks – 2500 companies;
- the ownership and financing is many-facetted.

There is no single dominating actor even though the universities play an important role. The innovation platforms have eight themes: the Environment Academy, Logistics, the IT Academy, the Food Network, Medicon Valley Academy, Nano Øresund, Diginet Øresund and the Entrepreneurship Academy. The Øresund Environment Academy is a non-profit network organization, working to strengthen the Øresund Region's science, business and education, with sustainable energy and environment as key areas. One main ØSR activity is promotion of the Øresund Region as an attractive place to invest and live.

Table 2.1 Conceptual comparison of the goals, methods and outcomes in four sustainable innovation case studies

	Science region	Leadership training	Mobility of experts	Hardwood system
Focus	Linking regional actors, businesses and universities	Leadership training for the company's executives	Enhancement of expert competence utilization	Development of supply chain cooperation
Process owner	Region authorities & 12 universities	The company itself and EFQM	Vinnova project and regional universities	Companies in a Vinnova project
Business development method	Promoting cross-border Triple-Helix partnerships, innovation & branding the region	Bench-marking by competent ambitious external reviewers, as a basis for open dialogue	Networking and dialogue, linking eco-design experts with regional business people	Advancement of a common value ground as a basis for cooperation and open dialogue
Investment dimension	Scientific networks and cross-border capacity building	Personal development for top-talent executives	Knowledge about access to advanced knowledge	Production development and business collaboration
Time frame	An ongoing long-term process of collaborative developments	Swift personal motivation effect, years for the wanted major explicit results	Numerous contacts, presentations and dialogues, during 1 year project	Explicit business development and closer contacts, within 2 years
Research method	Action research as a driving partner in networking	Action research as process leader for internal training	Action research as project leader for networking	Action research, promotion of competent dialogue
Innovation	More innovative collaboration – as measured by EU review. Visible as innovation center	Improved potential, but it is difficult to measure the specific business development effect	Hardy any visible innovation or direct explicit result, nice meetings is not enough	The companies have started more joint development, but it is not clear how innovative
Triple Helix	Double, one in each country	No, hardly relevant within a company	Yes, as a basic form of project idea	Not explicitly
Empowerment	Two spheres to relate to, aiming for tolerance, trust and flexibility	Yes, the hard review works as a form of confirmation as an interesting person	Several participants felt a form of confirmation during the meetings "Know who"	To some extent, the closer more explicit dialogue promotes self-confidence
Collaboration	Cross-border networks	Within and between companies	Probably not much short term effect – few knowledge-contacts were really mobilized	Explicit business development
Sustainable development effect	More openness for wider systems thinking & more chances to be able to test new ideas	Yes – EFQM goals include sustainability aspects. Empowered people can achieve more		It can be environmentally better to use Swedish hardwood than most alternative materials

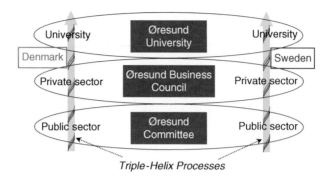

Fig. 2.6 The Øresund science region double triple helix

ØSR has developed an Innovation Guide (ØSR, 2008b) which describes *How and where to get support when commercializing your idea in the Øresund Region.* It introduces the region's innovation actors, technology transfer units and science parks. The guide presents the Danish and Swedish regional innovation systems and provides start up advice for innovators.

In 2008 ØSR was honored with a European Commission RegioStars Award for Regional Innovative Projects, in the category *Support clusters and business networks* (EC, 2008). These awards are intended to improve information about good practice, stimulate the exchange of experience and provide visibility for progressive-thinking. The jury's motivation for the ØSR award was

> This project has had real impact and has overcome language, legislative and physical barriers by building inter-regional partnerships that had previously been non-existent. It is an excellent example of a bottom-up triple helix approach. The project has a good administrative foundation, good working partnerships and is a good example of successful networking.
>
> (EC, 2008)

This is a positive grading which makes those working with ØSR both proud and empowered. The earlier Ideon related activities resulted in more advanced business development activities than if there had been no Ideon, in Lund. ØSR has most likely started to enhance the regions sustainability potential, but that is difficult to prove scientifically. Using the analogy with $R_{1,2}$ in Fig. 2.4, the question is if the activities are enhancing the ability to handle the challenges and capture the opportunities of tomorrow.

2.3.2 Leadership Training as a Driver in Sustainable Business Development

Djupenström (2008) has evaluated the European Foundation for Quality Management's (EFQM) leadership capability enhancement program, *Pegasus*, as a tool for sustainable business development. The leadership training pilot study was made at the Stora Enso Fors Mill in Dalarna, Sweden.

Among EFQM members, some persons were identified as "top talents" in their companies' talent management programs and were then assigned to take part in the Pegasus project for Leadership Capability Training. The training and assessment were facilitated by hiring an external EFQM consultant. The Pegasus participants first received 2 days training in various strategy tools and the EFQM Excellence Model. Both Pegasus participants and people representing the host (Fors Mill) participated in the preparation process to make sure that the Pegasus assessors had the needed information and material. The material they requested from the host included general information about the company and the business environment, as well as the prospective plans and strategy, especially considering the challenges that the company is facing. A site visit was planned and dialogue ensued with the Pegasus team and coordinators at Fors Mill.

During the site visit, employees at Fors Mill were interviewed by Pegasus assessors. The dialogues were based on the assessors' previous business experience – which resulted in creative ideas. Pegasus assessors' were guided by Fors Mill's coordinators and assigned guides. The assessment lasted 2 days and the Pegasus team used 1 day to summarize the findings. On the forth day, a workshop was held concerning overall perspectives and results together with the management team at Fors Mill and the Pegasus assessors. It included presentations of the findings and discussion concerning new ways of thinking.

The feedback from the process was appreciated as highly relevant by managers at the Fors Mill. The resulting suggestions for improvement were considered more innovative than earlier suggestions discussed in internal improvement sessions. The mentioned probable reasons are the multifaceted backgrounds and advanced experiences among the group of external assessors. The leadership evaluation process is considered to be a potent driver for innovative development.

EFQM's Fundamental Concepts of Excellence are:

- Results Orientation
- Customer Focus
- Leadership and Constancy of Purpose
- Management by Processes and Facts
- People Development and Involvement
- Continuous Learning, Innovation and Improvement
- Partnership Development
- Corporate Social Responsibility

Most of the EFQM criteria and the priorities in the Pegasus leadership training have some relationship to sustainable development priorities. The Fors evaluation resulted in 18 suggested improvement areas, whereof 16 were related to sustainable development priorities.

2.3.3 Mobility of Experts for Promotion of Regional Development

The Mobility Pilot Project was an action research and regional development project funded by the Swedish Governmental Agency for Innovation Systems, Vinnova (Karlsson, Bergeå, Berg, Pohl, & Kullin, 2003). It promoted mobility of experts in the fields of clean technology and health and safety. The experts came from universities in Dalarna, Kalmar and Luleå, three industrial R&D institutes (IVL, IVF, and SP), as well as the Swedish Trade Council. The goal was to enhance the renewal oriented ability of the regional companies and innovation systems, in the regions of Dalarna and Kalmar. One objective was to enable persons to act as sustainability oriented change agents.

The project start-up activities included lecturers from renowned companies with experiences from renewal activities. The main activities included local seminars in which local business leaders and regional development actors discussed with experts from academia, competence centers and progressive companies. The dialogues aimed to achieve mutual understanding, conceptual clarification and networking, between local SMEs, regional universities and national competences centers. The research dealt with transformative learning through interdisciplinary dialogue among persons with diverse knowledge and experiences.

The background included observations that, information exchange networks are important in establishing an environmental mentality in companies (Ehrenfeld & Lennox, 1997). One underlying hypothesis was that there is a need for a platform to establish a worthwhile environmental dialogue (Schlatter, 1998), in particular for interdisciplinary sustainability matters.

The project results indicate that mobility of environmental expertise, by itself, hardly has any direct explicit short-term effect in regional business development. It is difficult to judge if the project resulted in an adequate development of regional innovation systems, when compared to the 2000 person-hours of meeting time, during the project. However, the project team learnt a fair bit about the potential and challenges for sustainability oriented mobility and dialogue processes (Karlsson, Bergeå, & Luttropp, 2004; Karlsson, Magnus, & Huisingh, 2006; Karlsson et al., 2003). For example, the project experiences was used in the development of a PhD course on *Environmental efforts in competitive business development*, which investigated the dilemmas that proactive companies are struggling with (Bergeå, Karlsson, Hedlund-Åström, Jacobsson, & Luttropp, 2006).

2.3.4 Hard-Wood Business Development as a Driver for Sustainable Forestry

The forestry sector provides multifaceted values; these include the production of renewable raw-materials, sequestering carbon dioxide, as well as creating spaces for biodiversity, resilience and attractiveness as a place for recreation and as a landscape that humans appreciate. In many Swedish forests, hardwood trees support biodiversity, aesthetics and resilience (Karlsson et al., 2006).

Sustainable forestry is attracting increasing attention and in Sweden this includes an interest in nature conservation, and the incorporation of a diversity of broad-leaved trees. However, Sweden has a long tradition of major investments in coniferous forestry and expansion of softwood industries. There are 19% hardwood trees in the Southern Swedish forests, but for a number of years the market for Swedish hardwood timber has been low.

The presented studies started 5 years ago and included networking and open interviews with entrepreneurs and forestry related persons. Later on, a number of student projects have been conducted during the author's Ecodesign and Industrial Economics Courses. The perspective has a basis in systems thinking for learning organizations (Senge, 1990), including conceptual gaps between avoidance of risk and promotion of activity (Karlsson, 1998), e.g., between academic experts and employees in forestry related companies (Karlsson et al., 2004). The dialogues with students and their reports provided fresh eyes on the forest based business system. Renewal oriented seminars were organized with scientists, business persons, network actors and forest land-owners. Vinnova is now funding a project worth one million Euros for business developments to enhance the utilization of birch.

Three ecodesign students conducted twelve interviews, with experienced persons in the Swedish forestry and wood sector (Karlsson, Haug, Sjöberg, & Andersson, 2007). The dialogues aimed to promote sustainable forestry, entrepreneurship and human empowerment. The interviews were preformed as open dialogues, aiming for qualitative understanding. The method is inspired by quantitative interview methods, e.g., for the educational sciences (Kvale, 1997) and case study methodology (Yin, 1994).

One Vinnoa project ambition is to create higher added values for hard-wood through supply chain developments. The studies indicate that there is a need for concerted involvement of architects, designers, retailers, companies, municipalities and forest landowners, together with expertise from forest-based businesses (Karlsson et al., 2007).

The nature conservation interest often aims to prohibit logging of hardwood trees, as a safeguard against industrial exploitation, see Fig. 2.7. Contradictions between nature conservation and production interests are common. Such obstacles against collaboration have been researched, e.g., by Dryzek (1997).

To motivate more forest landowners to invest in cultivation of hardwood trees, it is important that there is a market interest in hardwood timber. To avoid that too many small hardwood trees are cut during thinning, it is vital to encourage business developments that correlate with promoting diversity in forestry. If there is no business interest in hardwood timber the hardwood seedlings are likely to be cut or mismanaged. The dialogues show that hardwood forest owners would appreciate a conceptual bridging between production and nature conservation interests, see Fig. 2.7 (Karlsson, 2007).

Fig. 2.7 A tree plant has potential as a renewable raw material producer and also as a part of the nature we want and need

2.4 Summary of Case Study Results

The goal of this chapter is to build conceptual understanding regarding the interrelation between different factors that influence the possibility to achieve a sustainable innovation process.

Table 2.1 presents a comparison of four different case studies in 11 dimensions. The following discussion includes comments on project economics, the driver metaphor and the investment concept, freedom of action, sustainable innovation as a core business priority, Triple-Helix and collaboration.

2.5 Discussion

The economic dimension is important. The expenses are obvious; however, on the contrary it is difficult to gauge the positive, often indirect and long-term, effects of renewal oriented investments. The connection between these elements becomes clearer when those who are positively interested also carry the costs. In this respect, the leadership training is a straightforward company internal process. The Hardwood System project is connected to the companies' business development interests and its project management and research is funded by Vinnova. As such, dialogue on expert opinions and suggestions is also deemed critical. The Mobility of Experts project has a regional development ambition and was funded by the Swedish Governmental Agency for Innovation Systems Vinnova.

One reason for why the ØSR initiative has attracted so much interest from universities is that universities funding has changed. Over the past 25 years, public support has decreased, from around 95% to some 35% today. This has spurred universities to look for new sources of funding. This entailed a change in the type of research. More research now aims to attract more funding from companies through contracts or sponsorship. The ØSR, with its double Triple-Helix approach, has provided a support platform for this. Still it has been a challenge for the further development of the initiative to secure multi-annual financing. It was difficult to mobilize SMEs to participate in the ØSR and its platforms, as they have little time to devote to activities other than their production and sales. Within the environmental platform especially, the funding proved difficult and could only be overcome with economic benefits.

The building up of knowledge is a form of investment. However it is important to remember that the resulting value is dependent on the how the future situation evolves; see Fig. 2.4.

Empowerment and freedom of action are important to make effective use of motivation and competence. **The driver metaphor** is most obvious for people that work within a company with a clear business development goal, i.e., leadership training. It is also rather clear for people that are working with explicit business developments, such as the Supply Chain development in the Hardwood System Case. *The Øresund Environment Academy* aims to provide a set of instruments and promote different kinds of entrepreneurship.

In the Mobility Case, the experts had environmental knowledge. The persons from the companies and regional business actors aim for business growth. Presentations at meetings included employment of environmental knowledge as a value adding support for business development. However the limited amount of activity later on indicates that it was difficult to "sell" and make actual use of the knowledge. It seems as if the business development "drivers" found it difficult to utilize the environmental "instruments".

The engaged and knowledgeable dialogue about environmental considerations, in Mobility project was judged to be valuable by the participating persons. However, it is unclear if the environmentally based insights resulted in more and better regional innovation activities. Short-term business developments do not depict if the Mobility Pilot Project made any real difference. Still, at least the project team members learnt a fair amount about the potential and challenges for sustainability oriented mobility and dialogue processes.

It seems to be more effective to integrate renewal oriented initiatives in the companies' internal competence development (Leadership Training Case), or in a more explicit cooperation (Hardwood System Case), than it is to introduce them through dialogue in an open network. Still, it is important for the mobilization of internal sustainability oriented change interests that there is a "tension" from legislation, external awareness and renewal oriented societal actors.

Using an analogy with Fig. 2.7, if there is low interest in a person's (a "plant's") potential to grow, it is likely to be mismanaged and degrade. Growth of monoculture forests is a risk from a sustainability point of view. In an analogous way it is a sustainability risk to have a lot of monoculture thinking. The understanding of the importance of sustainability in terms of nurturing a diversity of tree seedlings may be useful as an analogy when thinking about the growth of "entrepreneur plants". It is easier to "harvest" if all plants are alike, but the resilience is much better when there is a rich diversity of different kinds of plants.

One indication from the Hardwood Case Study is that it is difficulty to deviate from the forest industry's business-as-usual pattern and that there are risks for conceptual entrapment. To promote multifaceted entrepreneurship people in the forestry related businesses demonstrate that there is a need for positive examples and better branding.

2.5.1 Investments in Renewal Oriented Abilities

Looking at Fig. 2.4, the valuation of $R_{1,2}$ may seem unproblematic, when thinking in monetary terms. When looking at a recyclable material, the resource value for System 2 deviates from the resource value for System 1. The value of a recovered scarce alloy may decrease when an alternative with a more abundant raw material base evolves. On the contrary, the value of a recovered material may increase when it becomes possible to use it for something more valuable than earlier possible.

In an analogous way, the value of a certain kind of knowledge may decrease or increase; dependent on how the usage situation evolves. The value of knowledge about a conventional environmental technology, like a flue gas filter, may decrease if the fuel or combustion is improved. The value of environmental knowledge may increase considerably when it becomes useful in a more integrated and proactive way. The Mobility of Experts' Case was aiming for such an increase, but it was difficult to achieve. To engage in new more advanced forms of business oriented deployment of environmental information, it is important that the impending users understand the new form of potential value, which the environmental information is starting to get. However, it is difficult to understand the real sustainability effect of a form of use that surpasses the boundaries of established experience.

We here use System 1 to represent "end-of-pipe" thinking and System 2 to represent a modern clean-tech paradigm, in which the environmental knowledge is used in a more effective way. Then the transformation from the System 1 to the System 2 perspective describes that the environmental knowledge is reinterpreted to be useful in the development of the core business activities. However, it is a challenge that the new ability to make more advanced use of the knowledge has to be developed before the environmental knowledge becomes really appreciated in that way. Investments in personal ability, like in the Leadership Training Case, and investments in more open border-crossing contacts, like in the Science Region Case seem to be rather unproblematic from this point of view. The more advanced dialogues can open up for new more advanced understanding. Still, it does not replace the need for more advanced and far-reaching systems thinking. When a number of "end-of-pipe" experts collaborate they can make considerable improvements of that technology. But, to influence the core of the production system the environmental expertise has to be involved in the relevant development processes.

2.5.2 Sustainable Innovation as a Core Business Priority

It is crucial to become involved in the central part of the company's business development activities.

> Ultimately, the objective is to move climate change from the periphery to the core of the company. ... climate change must move to the departments best equipped to handle it. It must diffuse from EH&S to the core of the company's functional competencies and, in the process, become an issue of strategic importance to the company. ... you need to tap into...the functional levels best equipped to handle it, such as research and development, engineering, manufacturing, operations, marketing, finance, strategic planning, and /or human resources.
>
> (Hoffman & Woody, 2008, pp. 55, 60–61)

Sustainability concerns should be mobilized as a driver for renewal.

> You need the tension of a very challenging goal. Inspirational goals call an organization to act beyond conventional boundaries... An easy goal fails to challenge the creative potential of the organization.
>
> (Craig Heinrich in Hoffman & Woody, 2008, p. 45)

One challenge in the integration of sustainability oriented information exchange is that companies want to keep their core development plans secret. The confidentiality requirements of participating companies have been difficult to reconcile with ØSR's ambition to be transparent.

2.5.3 Freedom of Action

One basis for ØSR is that the Swedish and Danish cultures and languages are fairly similar, but also have their differences. Furthermore, the Swedish and Danish university systems are similar, and in this regard, it can be suggested that the two nations may be conceptualized as two different "customers" competing for researchers, teachers and students.

There is a growing interest in social entrepreneurship, which has deeper ambitions than profit. In the book, *The Power of Unreasonable people* (Elkington & Hartigan, 2008, p. 197) Shaw is quoted, *The reasonable man adapts himself to the world; the unreasonable one persists in trying to adapt the world to himself. Therefore all progress depends on unreasonable man.* According to the authors; "unreasonable" entrepreneurs are driven by passion – they work hard in critical phases often almost without salary. Even family and friends may consider them "crazy". Despite social stigmas, it can be recognized that they have a high level of priorities and motivation.

This is the true joy in life, the being used for a purpose recognized by yourself as a mighty one; the being thoroughly worn out before you are thrown on the scrap heap; the being a force of Nature instead of being a feverish little clod of ailments and grievances complaining that the world will not devote itself to making you happy.

(Shaw, 1903)

Unreasonable people have great energy, but still they are dependent on collaboration. The Øresund platform facilitates an easier process for innovative persons to find "friends". It grants more freedom of action to be able to relate to alternative cultures, formality systems and to have alternative groups of powerful persons to team up with.

2.5.4 Triple Helix

The Triple Helix holds that the university and its role in society is enhanced by engaging in the translation of knowledge into use, with feedback into theorizing and the opening up of new research questions as a positive consequence of such engagement.

(Etzkowitz & Zhou, 2006, p. 81)

The Triple-Helix process can be driven by a university (Dzisah & Etzkowitz, 2009) or other partners. From business point of view, it is often argued that Triple-Helix processes should be driven by business development actors. The Triple Helix collaboration became popular in the 1990s and it is promoted by Vinnova and ØSR.

The Triple-Helix concept implies that a concerted initiative can gather transformative strength. Figure 2.1 is also interesting as a metaphor. For example, plant growth is dependent on water, nutrients and space. As such, the synergy between three upward helixes of activity, investments and learning can enhance the total strength of the development process.

Over time, established power structures tend to become rather conservative. A Triple-Helix structure may provide openings for "unreasonable" entrepreneurs. When the established business actors are not interested; academia and/or political actors may still be interested. However, over time, the influential persons in the three spheres also tend to develop a consensus, in particular in delimited settings. The Øresund Region "Double Triple-Helix" Model aims to combine two different dimensions of freedom for action.

The ØSR Double Triple-Helix (Fig. 2.6) is different from the Triple-Helix Twins as described by Etzkowitz and Zhou (2006). The ØSR work is conceptualized as collaboration between two similar Triple-Helix Processes. The Etzkowitz Twin Triple-Helix model suggests that the "ordinary" Triple-Helix, that focuses business growth, should be balanced by an additional sustainability supervising Triple-Helix Process, aiming for enhanced control of environmental and social considerations. The ØSR Model aims to integrate sustainability oriented knowledge in the cores of both their Triple-Helix Business Development Processes.

New support structures may open up for various nonconforming initiatives. However over time, also new support structures tend to be incorporated within already established structures. Consequently it is important that there are individuals and business angels that trust and dare to act on their intuitive appreciation of renewal oriented opportunities.

Real life observations are important as drivers for transformative learning. Textbooks and also scientific thinking tend to focus on one dimension at a time. Open-minded observations of reality may provide more eye-opening "truth" than textbook theories, "The devil is in the details". However, no observation can be independent of the earlier understanding. New theories can enable people to discover and understand new aspects.

Dialogues between people that have different kinds of experience are important to broaden traditional perspectives and thinking patterns. ØSR has a diversified base and its management structure aims to establish tolerance, trust and flexibility. One linkage between theory and practice is that the Øresund Entrepreneurship Academy combines practical experience with the receipt of academic credits. The Øresund Initiative has enabled "light institutionalization" of science/industry relationships in a cross-border region, thereby removing mental barriers to cross-border science/industry collaboration and opening collaborative opportunities.

Activity systems such as ØSR can enhance the quality of innovative activities and the learning process. Knowledge can improve each time it is used. Learning from real-life "experiments" can be "recovered" and "recycled" (analogous to $R_{1,2}$ in Fig. 2.4). The sustainability value of a specific knowledge may increase, from N to $R_{1,2}$.

However it is difficult to "measure" the sustainability value, as it has many dimensions. There is a need for a more structured way of thinking, as a base for dialogue. To mobilize creative talent it is vital to promote freedom of action and multifaceted creative dynamic thinking. To make new knowledge effective in innovative business development it is essential to generate concerted action. Triple-Helix processes have potential to combine these ambitions.

2.5.5 Collaboration

To enable collaboration between sustainability-oriented change agents, with different perspectives, it is essential to generate common understanding and mutual trust. The motivation for concerted action is dependent on the social capital:

> The term social capital captures the idea that social bonds and norms are critical for sustainability. Where social capital is high in formalized groups, people have the confidence to invest in collective activities, knowing that others will do so too.
>
> (Pretty, 2003, p. 1912)

The most common formalized group is a company, with the Leadership Training Case representing a company's internal process. The EFQM training resulted in stronger social bonds and common norms among the participating persons. The motivation to collaborate improved.

To achieve a real result, the innovative process has to have strength and freedom of action. It is difficult to define criteria for sustainable innovation. Still it is necessary to try. One basis is that sustainability is multidimensional; economic, environmental and social. Furthermore, it is important to relate to life-cycle thinking and scientific knowledge. However, it is also important to recognize that scientific knowledge changes over time. In an analogous way as the valuation differs between the systems in Fig. 2.4, the sustainability relevance of certain knowledge may change.

The sustainable development discourse entails a need for change, i.e., relearning through a more effective feed-back (Senge, 1990). Dialogues that convey honest lucid responses to each other's ideas may provoke a process of rethinking and promote a renewal oriented personal development. This aspect seems to have been most effective in the Leadership Training Case. The critical assessment, combined with straightforward feed-back from external assessors, granted an effectual form of feed-back. The explicit business developments in the Hardwood Supply Chain have also resulted in thought-provoking enlightenment.

The ØSR Initiative demonstrates that active communication is vital to help innovation support systems become more transparent. Tolerance for the two languages on equal footing, without the use of translation or interpretation, is important for the dialogue. ØSR also suggests that innovation support is effective only when it is provided by professionals. Still, relaxed dialogues and access to expertise, as in the Mobility of Experts case, is not sufficient.

To accomplish a resilient process of change there is a need for transformative learning (Mezirow, 2000). "In order to effectively guide people and organizations in transformative learning, one needs to reflect on experiences, theories and on epistemic questions" (Bergeå et al., 2006, p. 1441). Long-lasting change of people's thinking and acting will never happen unless they want to change and participate in sustainable development processes.

> In a culture that perceives competition – not collaboration – as the great animator, it makes sense that taking "a critical stance" and defining "the very best" argument would be the procedure of choice.
> That procedure, however, effectively shuts out immature or marginalized people. Critical discourse, the doubting game, can only be played well on a level playing field. The believing game, in contrast, is the game for everyone no matter how immature or silenced.
> (Mezirow, 2000, p. 89)

To promote an Agenda 21 process of change; Lindström and Johnsson (2003) suggest active participation and feedback. In accordance with Thomas and Velthouse (1990) they define empowerment as "a cognitive state that results in intrinsic task motivation" (Lindström & Johnsson, 2003, p. 21).

On a personal level the "intrinsic task motivation" is similar to the state of mind that Csikszentmihaly (2002) describes as "Flow", the positive synergy between being excited and having a feeling of control. Looking at special education, Karlsson and Westerlind (2006) suggests that self-esteem is essential to enable a positive feedback loop between the teacher and the student; see also Burn (1979). One basis for transformative learning is to have a sufficient self-esteem to feel safe enough to really try to advance one's thinking and activities beyond what one has been thinking and doing before.

To support self-esteem, collaborating individuals ought to show appreciation of each other as persons. One basis for trans-disciplinary learning is that the group can detect and understand many different kinds of observations and viewpoints. The collaborating group ought to include a variety of persons in an open dialogue in which they try to elucidate their different thoughts and observations to each other.

The Table 2.1 summary illustrates that it is a multifaceted issue to build sustainability oriented collaboration. The Leadership and Hardwood cases focus on explicit personal training and business development, and those processes are found to promote personal development as well as the enhancement of the collaborative competence. The Science and Mobility cases have more focus on the networking itself and this appears to enhance the level of sustainability-oriented understanding. The Øresund Science Region Triple-Helix process aims to mobilize sustainability knowledge in more innovative collaboration.

2.6 Concluding Remarks

The sustainability challenges imply that there are severe needs for intelligent innovation processes. This chapter presented four case studies on sustainability oriented collaboration; including experiences from advanced leadership training, the Øresund Science Region innovation system, mobility of sustainability expertise, as well as business developments for hardwood. The chapter highlights the importance of supporting entrepreneurship and transformative learning to deliver intelligent innovation. The entrepreneur is presented as a "driver" with a need for appropriate instruments, a steering wheel and an inspiring vision. In terms of transformative learning, self-esteem, empowerment and freedom of action are essential. The feedback learning effect is improved when a group of persons are truly engaged in trying to understand and do something genuine together. In addition to open dialogue, it is vital to mobilize sufficient concerted "investments" in innovative sustainable development "experiments". To build motivation to engage in radical renewal activities, it is important to exchange ideas with stakeholders that can understand the new form of potential value, that that entrepreneur is aiming for. This can be facilitated by open-minded Triple-Helix collaboration.

References

Bergeå, O., Karlsson, R., Hedlund-Åström, A., Jacobsson, P., & Luttropp, C. (2006). Education for sustainability as a transformative learning process: A pedagogical experiment in EcoDesign doctoral education, In Special issue on *EcoDesign – What's happening?. Journal of Cleaner Production, 15*, 1421–1442.

Burn, R. B. (1979). *The self-concept in theory, measurement, development, and behaviour.* London: Longman.

Carpenter, J. (2008). Metaphors in qualitative research: Shedding light or casting shadows. *Research in Nursing and Health, 31*(3), 274–282.

Csikszentmihaly, M. (2002). *Flow – The classic work on how to achieve happiness.* London: Rider.

Djupenström, A. K. (2008). *Business excellence models as a driver for sustainable development.* Ecodesign Course Thesis, University of Kalmar, Kalmar, Sweden.

Dzisah, J., & Etzkowitz, H. (2009). *Triple helix circulation: The heart of innovation and development.* http://www.triple-helix-7.org/theme-paper.htm , January 30, 2009.

Dryzek, J. (1997). *The politics of earth: Environmental discourses.* New York: Oxford University Press.

EC. (2008). European commission. *Regions for economic change – Exchanging good practice between Europe's regions.* http://ec.europa.eu/regional_policy/cooperation/interregional/ecochange, June 6, 2008.

Ehrenfeld, J., & Lenox, M. (1997). The development and implementation of DFE programs. *Journal of Sustainable Product Design, 8,* 17–27.

Elkington, J., & Hartigan, P. (2008). *The power of unreasonable people – How social entrepreneurs crate markets that change the world.* Boston, MA: Harvard Business Press.

Etzkowitz, H., & Zhou, C. (2006). Triple helix twins: Innovation and sustainability. *Science and Public Policy, 33*(1), 77–83.

Hoffman, A., & Woody, J. (2008). *Climate change – What's your business strategy?, Memo to the CEO.* Boston, MA: Harvard Business Press.

Karlsson, R. (1994). *LCA as a guide for the improvement of recycling.* Proceedings of the European workshop on allocation in LCA, CML, Leiden, 24–25 February, SETAC, Brussels, The Netherlands.

Karlsson, R. (1998). *Life cycle considerations in sustainable business development, eco-efficiency studies in Swedish industries.* PhD thesis, Chalmers University of Technology, Göteborg, Sweden.

Karlsson, R. (2007). *Business diversity and sustainability reporting as proactive agents for sustainable forestry.* International Scientific Conference on Hardwood Processing, 24–26 September 2007, Quebec City, Canada.

Karlsson, R., Bergeå, A., Berg, P., Pohl, E., & Kullin, J. (2003). *Rörlighet hos nyckelpersoner som verktyg i affärsutveckling för hållbar tillväxt.* Report 2003-00248, VINNOVA, Stockholm, Sweden.

Karlsson, R., Bergeå, A., & Luttropp, C. (2004). *Mobility of key persons for promotion of business renewal and sustainable growth.* The 3rd Int. Conf. on Systems Thinking in Management (ICSTM 2004), 19–21 May 2004, Philadelphia, PA, USA.

Karlsson, R., Haug, A., Sjöberg, A., & Andersson, P. (2007). *Academia's role in clarification of linguistic and systemic relations between wood business developments and sustainable forestry.* The 13th Int. Sustainable Development Research Conference: Critical Perspectives on Health, Climate Change and Corporate Responsibility, 10–12 June 2007, Västerås, Sweden.

Karlsson, R., & Luttropp C. (2006). EcoDesign: what's happening? An overview of the subject area of EcoDesign and of the papers in this special issue, Editorial in Special issue on *EcoDesign – What's happening?. Journal of Cleaner Production, 14*(15–16), 1291–1298.

Karlsson, R., Magnus, L., & Huisingh, D. (2006). *Product design as a key to a business system perspective that promotes sustainable forestry.* The 50th Annual Meeting of the International Society for the Systems Sciences Complexity, Democracy and Sustainability, 9–14 July 2006, Sonoma State University, Rohnert Park, CA, USA.

Karlsson, G., & Westerlind, G. (2006). *Självkänsla och specialpedagogiskt stöd* (Self-esteem and special education), C-uppsats i specialpedagogik; Högskolan i Kristianstad.

Kvale, S. (1997). *Den kvalitativa forskningsintervjun.* Studentlitteratur, Lund, Sweden.

Lindström, M. (2003). *Attitudes towards sustainable development. Priorities, responsibility, empowerment,* PhD Thesis, Lund University.

Lindström, M., & Johnsson, P. (2003). Environmental concern, self-concept and defence style: A study of the Agenda 21 process in a Swedish municipality. *Environmental Education Research, 9*(1), 51–66.

Meisser, S., & Barry, J. (2007). Through the looking glass of organizational theatre: Analogically mediated inquiry in organizations. *Organisational Studies, 28*(12), 1805–1827.

Mezirow, J. (2000). *Learning as transformation*. San Francisco, CA: Jossey-Bass.

Morgan, G. (1986). *Images of organisation*. Thousand Oaks, CA: Sage Publications.

Murphy, R. (1994). *Rationality and nature, A sociological inquiry into a changing relationship*. Boulder, CO: Westview Press.

Ortony, A. (1975). Why metaphors are necessary and not just nice. *Educational Theory, 25*(1), 45–53.

Pretty, J. (2003). Social capital and the collective management of resources. *Science, Special Section on "Tragedy of the Commons?" , 302*, 1912–1914.

Schlatter, A. (1998). *Umwelt-Dialog – Ökologieorientierte Lernprozesse in Unternehmen* (Diss. ETH Nr. 12708). Zürich: ETH Zürich.

Senge, P. (1990). *The fifth discipline: The art and practice of the learning organization*. New York: Doubleday/Currency cop.

Shaw, G. B. (1903). One True Joy in Life, in the four act drama *Man and Superman – a Comedy and a Philosophy*. Westminster Archibald Constable and Co http://www.quotationspage.com/quote/27168.html, June 6, 2009.

The Norwegian Nobel Committee. (2006). *Nobel Prize Speech*. Retrieved May 28, from http://nobelprize.org/nobel_prizes/peace/laureates/2006/presentation-speech.html.

Thomas, K. W., & Velthouse, B. A. (1990). Cognitive elements of empowerment: An "interpretive" model of intrinsic task motivation. *Academy of Management Review, 15*, 666–681.

VINNOVA. (2002). *Effective innovation systems and problem-oriented research for sustainable growth, VINNOVA's strategic plan 2003–2007*, VINNOVA, VP 2002:04, Stockholm, Sweden. www.vinnova.se/In-English/Publications/VINNOVA-Policy , June 6, 2008.

WCED. (1987). *Our Common Future* (The Brundtland Report) World Commission on Environment and Development. London: Oxford University Press.   Chapter 2 :1:15 The Concept of Sustainable Development.   http://www.un-documents.net/ocf-02.htm , June 15, 2009.

Wheatley, M. (2002). *Turning to one another – Simple conversations to restore the hope to the future*. San Francisco, CA: Berrett-Koehler Publishers.

Yin, R. K. (1994). *Case study research: Design and methods* (2nd ed.). Thousand Oaks, CA: Sage Publications.

ØSR. (2008a). Øresund science region. http://www.oresundscienceregion.org, June 6, 2008.

ØSR. (2008b). Øresund science region. *Innovation Guide 2007*. http://www.uni.oresund.org/Oeresundsuniversitet/Dokumenter, June 10, 2008.

Chapter 3
Partnerships and Sustainable Regional Innovation Systems: Special Roles for Universities?

Martin Lehmann, Per Christensen, and Björn Johnson

Abstract The notion of Public–Private Partnerships (P3) is ambiguous. To date, however, there has been little emphasis on universities in this connection, and their roles (if any) are still somewhat unclear. The question we ask, therefore, is: What is or could be the role of universities in P3s? In this chapter, the first part is dedicated to the discussion and clarification of the concept of public–private partnerships. The role of universities if and when actively participating in 'life outside the ivory tower' is addressed. These partnerships are also discussed in a regional context. With the point of departure in innovation theory, we combine 'sustainable development' with the Regional System of Innovation approach to propose a new concept – Sustainable Regional Innovation System – in which regional initiatives such as Public–Private–(Academic) Partnerships play an integrated role, not least in the context of 'learning and innovation for sustainable development'. Two cases are presented to underline the importance of what is signified as Public–Private–Academic Partnerships (PPAP); i.e., partnerships, where universities are given – or take on themselves – a specific role. In such partnerships, we argue, mediation is a major function of universities, including both the provision of new knowledge and the conciliation of opposing views, and universities thus act as catalytic and institutionalizing entities.

Keywords Public–private partnerships · Triple helix · Regional sustainability · Innovation · Case studies

3.1 Introduction

Partnerships emerge through mutual trust and commitment and as a result of social relationships and power relations. In terms of sustainable development, partnerships,

M. Lehmann (✉)
Department of Development and Planning, Aalborg University, Fibigerstraede 13, Aalborg E, 9220, Denmark
e-mail: martinl@plan.aau.dk

J. Sarkis (eds.), *Facilitating Sustainable Innovation through Collaboration*, DOI 10.1007/978-90-481-3159-4_3, © Springer Science+Business Media B.V. 2010

especially Public–Private Partnerships (P3), should be one of the new pivotal mechanisms of greening. The notion underpins the shift in regulatory regimes that, through political and ecological modernization, has been going on for more than a decade. The World Summit on Sustainable Development (WSSD) in Johannesburg in 2002 actively promoted the establishment of such partnerships, which, on the one side, should revolve around sustainable development as a goal, and on the other, the voluntary collaboration between communities, governments, businesses and NGOs to achieve this goal.

The diversity and range of scholarship in the field is considerable and includes, for example, studies on partnerships in the US prison system (Schneider, 1999), global partnerships in health and for health development (e.g., Bazzoli et al., 1997; Buse & Walt, 2000), partnerships for urban governance (e.g., Pierre, 1998), partnerships for environmental management (e.g., Glasbergen, 1998, 1999; Manring, 2007), and partnerships for sustainable development (e.g., von Malmborg, 2003; Roome, 2001).

The term P3 is general, it is applied to a number of different subjects, and the partnerships are formed for a multitude of reasons. Some partnerships are local, others national or regional, and some are even international. Further, some collaborations are corporatist arrangements while others produce a set of relations that may be termed non-market interaction (Glaeser, 2000; Glaeser & Scheinkman, 2001; Sjöberg, 1993; Sorensen, 1994). This indicates a qualitative distinction from market relations and corporate arrangements and raises the issue of why such relationships are established, maintained and developed (Lehmann, 2008).

In environmental services, for example, corporate-type collaborations may be found in the privatization and operation of water and sewage works, and wastewater treatment plants, with goal to provide equal or better environmental service while maintaining economic feasibility. Often of a local or regional nature, these partnerships may also be supported by the international community through international organizations' programs. The United Nations Development Program initiative 'Public–Private Partnerships for the Urban Environment' is one such program. In this context, the partnership can be viewed as involving contractual obligations and relations, as well as transfer of responsibility.

Other definitions of P3 focus more on collaborative aspects and the formation of partnerships as a new form of cross-sector collaboration or as a network between several parties that have common objectives and are united in achieving their goals. In this context, The Copenhagen Centre (which itself can be defined as a public–private partnership) provides a meaningful, albeit broad, definition:

> People and organizations from some combination of public, business, and civil constituencies, who engage in voluntary, mutually beneficial, innovative relationships to address common societal aims through combining their resources and competencies. (Nelson & Zadek, 2000, p. 14)

This definition of partnership is that of a 'social partnership', i.e., focusing on aspects of social cohesion and economic competitiveness. However, the definition can be equally valid in a broader sustainability context. It can also be used in the less broad environmental context, where economic competitiveness is of equal importance, while the notion of social cohesion is replaced by that of environmental management. The contents and principles of the partnership will differ, however, and as a consequence, so will the success criteria.

Often, the major stakeholders in public–private partnerships are from government (local, regional, national), non-governmental organizations, international organizations, and private companies (LaFrance & Lehmann, 2005). All have their own particular reasons for joining or initiating a partnership, with each contributing varying competencies and resources. It follows that multi-stakeholder partnerships thus consist of more than two major stakeholder-groups, and may in fact be seen as a new form of governance (Lehmann, 2008).

With universities being increasingly recognized as playing important roles in achieving sustainable development, it may make sense to distinguish between partnerships without and partnerships with strong academic involvement. Underlining this is the fact that academic institutions bring particular resources to the table in a partnership (Hansen & Lehmann, 2006), and at the same time can benefit in terms of research and education from the closeness to other actors (Christensen, Thrane, Jørgensen, & Lehmann, 2009; Lehmann, 2008; Lehmann, Christensen, Du, & Thrane, 2008). Mutuality is thus present. The notion Public–Private–Academic Partnerships (PPAP), which may also be referred to as 'triple-helix partnerships' may then be more suitable to cover the latter activities, and the notion of P3 should be left to activities where academia is not directly and explicitly present. Both may still be looked upon as forms of governance.

However, academia can also play an important role in innovation systems. Universities are increasingly recognized as key actors in national innovation systems (Edquist, 2005; Mowery & Sampat, 2005). In fact, in many innovation studies there is a strong focus on high-tech, science-based innovation and on the interactions between big firms, universities and other research organizations. But universities are not only important in innovation systems dominated by science-based production. There are many different connections between universities and the societies in which they operate. They provide firms with employees with science-based educations; they produce new scientific knowledge, which firms can use; and they cooperate with universities in research and in other ways as well. Universities also function as mediators and translators in such partnerships (Hansen & Lehmann, 2006).

Recently, the so-called third mission of universities (in addition to research and teaching) has drawn attention to a diverse and broad set of relations between universities and society. Third mission activities are concerned with the generation, use, application and exploitation of knowledge and other university capabilities outside academic environments (Molas-Gallart, Salter, Patel, Scott, & Duran, 2002). Also the so-called Mode-2 type of knowledge production leads to new relations and closer interaction between universities and society (Nowotny, Scott, & Gibbons, 2001).

This situation gives rise to the discussion of whether PPAP should be seen as regional initiatives and from a networking/governance point of view, or whether they can be better described, understood and utilized through an innovation system approach. In this chapter, we use the latter approach in order to clarify and discuss triple-helix type partnerships with special focus on regional innovation systems in relation to sustainable development. Section 3.2 provides a short overview of the systems of innovation approach and the framework for partnerships in this light. This is followed by a presentation (Section 3.3) and discussion (Section 3.4) of two cases of partnerships in different regions of Denmark. Finally, conclusions are presented, outlining these collaborations as being in fact *both* governance and part of innovation systems.

3.2 Systems of Innovation

The concept of 'systems of innovation' emphasizes the interdependence and interaction between technical and institutional change in the process of development. The main idea behind the concept is that the innovation performance of an economy (nation, region, city) depends not only on how its individual firms and organizations perform, but also on how they cope with change and interact with each other and with the financial and public sectors.

There are narrow and broad versions of the innovation systems approach to economic dynamics. In the narrow approach, the focus is on the research and development system and on high-tech activities and science-based production. In the broad version, innovations are also seen as anchored in everyday activities like procurement, production and marketing in all kinds of firms, organizations and sectors, so that innovation includes small, incremental improvements of processes and products as is also often found in environmental public–private partnerships.

Within the broad conceptualization of innovation systems, there are at least three propositions. First, specialization in terms of production, trade and knowledge is important for innovative performance. The focus is on the co-evolution between what countries and regions do and what people and firms in these countries and regions know how to do well. This proposition implies that both the production structure and the knowledge structure will change only slowly, and that such change involves learning.

Second, some of the elements of knowledge are localized and not easily moved from one place to another. A central assumption behind the innovation system perspective is that knowledge is more complex than information, and that it is not always codified or even possible to codify but also includes tacit elements (Polanyi, 1966). Important elements of knowledge are embodied in the minds and bodies of agents, in the routines of firms and, not least in the relationships between people and organizations (Dosi, 1999). This makes knowledge spatially sticky (von Hippel, 1994) so that to some extent it adheres to the place where it was created.

Third, relationships and interactions between people and organizations matter. The relationships serve as carriers of knowledge, and the interactions as the process by which new knowledge is produced and learned. This assumption reflects the fact that neither firms and knowledge organizations nor people innovate alone. The crucial point is that interactions between people and organizations have the potential to combine different kinds of knowledge, insights and competences in new ways and that this supports innovation (Jensen, Johnson, Lorenz, & Lundvall, 2007).

Characteristics of interaction and relationships may be called 'institutions'; institutions are informal and formal norms and rules regulating how people interact (Edquist & Johnson, 1997; Johnson, 1992; Scott, 2001). The institutional approach implies that history and context make a difference when it comes to how agents interact, learn and innovate. An understanding of innovation processes is not possible without at least some grasp of how institutions shape interactive learning and innovation.

3.2.1 Regional Innovation Systems

The broad approach to systems of innovation argues that there are many possibilities and that there is no ideal territorial base where innovation will always flourish (see, for example, Edquist, 2005; Edquist & Johnson, 1997; Edquist & McKelvey, 2000; Johnson, 2007; Lundvall, Johnson, Lorenz, & Lundvall, 2002). First of all, we should look for a geographical area that shares institutional characteristics that lead to frequent, intense and high-quality interactions. We should also look for an area with a certain degree of production and trade specialization, namely an area where, over time, people and firms have become good at doing certain things and acquired a production and competence profile of some sort. Accumulated competence contributes to specific interaction characteristics for the area in question and impinges on the processes of innovation. Furthermore, we should consider areas with a common knowledge infrastructure (including, for example, schools, universities, research institutes, technological service centers), with governance structures and with some kind of public policy routine. This characteristics includes an established polity, with policies affecting learning, innovation and governance directly and indirectly. Finally, we should identify regions that have acquired specific demand characteristics that match its specialization pattern and enable different kinds of organizational interactions.

A spatial delimitation with all these characteristics is not easy to find. Small and reasonably culturally homogenous nation-states seem to be obvious candidates as 'national systems of innovation' (Lundvall et al., 2002). Many types of regions also have some or most of the characteristics identified above; hence, the lively research concerning regional systems of innovation (Asheim & Gertler, 2005; Cooke, 1992). Cities may also possess many of the characteristics which form a good innovation system, and the usefulness of the notion of city systems of innovation has also been proposed (Johnson, 2007; Johnson & Lehmann, 2006). Finally, a local community or a group of such communities may also constitute an interaction area and a local system of innovation is now increasingly proving to be a useful concept in development theory and policy (Cummings, 2005).

Regional systems of innovation can are the institutional infrastructure supporting innovation within the production structure of a region (Asheim & Gertler, 2005).

For our focus in this chapter, we can also include environmental management performance and governance. Such innovations are produced with the help of regional networks of innovators, regional clusters and industrial districts. As an interactive process innovation is very often regionally contained since it depends on combining tacit knowledge with codified knowledge and learning by doing, using and interacting with more science-based learning, which requires face-to-face contacts and trust-based relations. This environment is supported by the proximity between actors and the traits of common culture that sometimes exist within a region.

3.2.2 Regional Partnerships as Systems of Innovation

Hansen and colleagues (Hansen & Lehmann, 2006; Hansen, Lindegaard, & Lehmann, 2005) argue that universities must find their roles with regard to learning, education and research through an active coupling with practice. An example of this goal is delivering graduates and providing partnerships in research and education to business, civil society and government. A broad range of such third-mission activities tends to anchor universities more firmly in their regions involving a range of constituencies. van Kerkhoff and Lebel (2006) provide similar observations:

> (...) we reached a contrary view of the world, one in which research, politics, researchers and publics are intertwined in a constant struggle of justifications, explanations, and decisions in an uncertain and complex world. These questions encourage us to look at the *relationships between research-based knowledge and action as arenas of shared responsibility, embedded within larger systems of power and knowledge* that evolve and change over time. This conceptualization offers a more appropriate starting point for understanding the role of research in sustainable development than the conventional model of trickle-down, transfer and translation. (van Kerkhoff & Lebel, 2006, p. 473; emphasis added).

These shared arenas are examples of PPAP (Lehmann, 2008, p. 48):

> Stakeholders from some combination of public, private, academic and civil constituencies, who engage in voluntary, mutually beneficial, and innovative relationships to address and build natural, human & intellectual, production and social potentials through combining their resources and competencies.

Figure 3.1 shows how this relationship can be understood with an environmental focus, technology and collaborative projects and ending in governance and sustainability. The lighter triangle signifies approaches to sustainability-related problems, and the darker triangle and notions (a) through (c) signify approaches to PPAP and embedded actions.

The division of activities into three levels that mutually are non-exclusive signifies distinctions in work on different partnership levels in terms of time, commitment, member diversity, and ease of entering and/or leaving the partnership. Thus, the levels signify distinctive complexities in terms of (1) 'relationships', (2) 'actions' and (3) 'systems' (of power and knowledge). These levels are (in order of increasing complexity):

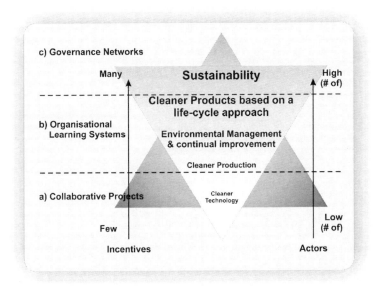

Fig. 3.1 The greening triangle and collaborative partnerships (darker, background triangle)

(a) collaborative projects
(b) organizational learning systems, and
(c) governance networks.

Collaborative projects are defined as limited in time, commitment of partners may be relatively small. Their range may be limited can be relatively easily replaced. At the second level are organizational learning systems. At this level the commitment is necessarily higher; partners are challenged and may as a result start 'doing business' differently, implementing changes in their respective organizations. While the diversity of partners may not necessarily be higher than in (a), the inter-dependency (Siebenhüner, 2005) is, and outcomes may be negatively affected should partners decide to leave. At the final level, governance networks are found. This is where institutional changes may occur and rules of the game begin to change, but they will not do so without very strong trust and commitment from a wide variety of partners, both in terms of developing new rules and in playing by them. Time-wise, this type of partnership requires more than the levels (a) and (b), and may in many cases be building on these.

3.3 Two Cases of Partnerships

In Denmark, several cases of quite similar partnership approaches exist. The first were formed in the early 1990s while others have more recently come into existence. In the following, two cases of these types of partnerships are presented.

The first is the Green Network, which is generally accepted as the inspiration and role-model for later networks. The second case is the Sustainable Business Forum North Denmark, which, while also inspired by the Green Network and utilizing many of the tools it developed, is different in the sense of which actors take an active and leading role and who initiated the network.

3.3.1 Green Network

The story of the Green Network begins in 1992 when the seeds were planted for the first environmental networks with strong and binding participation from both the public and the private sectors in Denmark. The then National Agency for Trade and Industry launched a competition with the aim of establishing a network that was to be the Danish showcase for environmental knowledge and technology. From that competition, one of the most successful networks in Denmark was established, namely the Green Network in the county of Vejle (Lehmann, 2006). It did not win the competition nor did it become an international showcase in the original sense (Lehmann, Christensen, & Larsen, 2005). Significant effort had gone into establishing the co-operation and the network, however, and neither the public sector agencies nor the private companies involved were prepared to just write that off. Instead, it reconstituted itself, making the network more local. The promotion of environmental activities was preferred instead of the previous focus on economy and exports (due to the competition guidelines; Green Network, 2003, with co-ordinator Erik Ørskov, Green Network Secretariat, Personal Communication). The Green Network became a P3 focused on demonstrating that – locally as well as regionally – environmental considerations and economic gains as well as the private sector and the public authorities could indeed collaborate and contribute to greater societal benefits. From the outset, and currently, the major drivers in the network are private companies and public authorities. This characterization is reflected both in the network's organization and in the activities taking place, showing that the benefits must come to those who commit resources.

The pivotal mechanism in the workings of the Green Network was to be a recognized, seminal form of 'green accounting' (statement) and environmental management system. This was developed through a collaborative project with participation from several local companies, the local municipalities and the Danish consultancy firm, COWI A/S. The concept revolved around a dialogue-based approach to local government's obligations of granting environmental permits and inspecting companies. Visual recognition is provided through the Green Network flag and diploma.

In June 1994, the Green Network was formally established with organization, by-laws, activities, and a business plan. A three-tiered membership was established reflecting both obligations and responsibilities toward the Network and its activities. Vejle County and the municipalities of Vejle, Horsens, Kolding, Fredericia and Middelfart constituted the public sector (O-members), approximately 30 companies formed the most active of the private sector (V-members), and a similar number of other organizations were part of the network as so-called Interested Parties (I-members). The economy of the network was and still is based partly on membership fees[1] and partly third-party funding for various projects.

The network has been able to move from an initial focus on environmental management to today's integrated approach on sustainability (Lehmann, 2006, 2008), and meanwhile it has also expanded its membership base and its outreach. Academia played a small part in this move through active involvement in various seminars and workshops calling for enlarging the focus beyond environment only. In 2001 and 2002, faculty from Aalborg University (a university located outside the Green Network's region) were commissioned to participate in the strategic reform of the network with the goal of formulating a new network strategy that should encourage and recognize organizations' work with occupational health and safety and social accountability alongside that of environmental management.

The strategic move from 'environment' to 'sustainable development' was a reality in early 2004 and not much later the Green Network toolbox of simple managements systems and 'back-pack' of supportive activities were enlarged in order to enact the new sustainable development strategy. A brief overview of activities based on Lehmann's (2006) categorization is presented in Table 3.1.

The Danish administrative structural reform,[2] which took place in 2005–2006 and took effect as of 1 January, 2007, shook the network's foundations slightly. The network had based its formation and its support functions on the county administration. Abolition of the county-level of government (including its strong economic support) could have meant the end of the Green Network. This, however, has not been the case.

Table 3.1 Overview of activities of the Green Network

Category	Activities
Statements	Process consultants (EMS, OHS, CSR), dialogue, Review & Certification (of statements),
Continuing education	Process consultants, short courses, seminars & workshops
Projects	Examples: Environmental management in agriculture, Green Purchases, LCA manual, social reporting, environmentally friendly transport, workers' health, toolbox on 'waste and re-use', guide on chemicals, establishment of Key2Green.
Information & communication	Newsletters, press releases, internet website, diploma events, speeches, yearly green account

[1] Until the Danish structural reform, this included the county's contribution towards the running of the network's secretariat.

[2] The local government reform created a new map of Denmark; 98 municipalities replaced the previous 271, the 13 counties were abolished and five new regions created, and a new division of tasks between local, regional and state authorities took place. Cf., e.g., http://www.sum.dk/publikationer/government_reform_in_brief/index.htm.

In southern Denmark, four networks are now active to varying degrees. The Green Network is the most active and has in fact expanded its administrative support functions in order to be able to honor the activity level of its members. The network is now based on six municipal O-members and approximately 260 member organizations, public as well as private.

The new regional authority, the Region of Southern Denmark, is increasingly showing an interest in using the networks' expertise, drive and commitment to expand sustainability initiatives geographically as well as conceptually. This interest is backed by a commitment of considerable human as well as institutional and economic resources, but does not constitute the same outright regional support as was the case before the reform (where the County of Vejle was both a founding member and directly supported the Green Network). Recent developments, however, include a permanent seat for the Region on the network's Board as well as participation in various co-ordination groups.

By way of this inclusion, the Region has pushed for stronger university involvement, not least by encouraging and supporting the networks and the university in the region to come together in a strategic effort to enlarge the networks' membership base, and in jointly developing new projects aimed at innovating both the networks' various toolboxes and the universities' related curricula. Some of these projects (total budget of approximately €2.5 million) are now being implemented while others have failed to secure funding for the proposed activities. Whereas these projects have enhanced collaboration with academia, it is not yet evident if the collaboration will be sustained.

The most recent development in this regard is an invitation extended to Aalborg University to establish a permanent base at the offices of the Green Network. According to the Network (Green Network, 2009, with Manager Dorthe Bramsen Clausen, Green Network Secretariat, personal communication), this presence should encourage and enable researchers and students of the university to work more closely with both the network members and support staff, and vice-versa.

This initiative is now emerging and will eventually show whether the Green Network will stay a P3 with academic involvement on a project-to-project basis only (and keep the strong focus on governance) or turn into a PPAP, and thereby include a stronger focus on innovation.

3.3.2 Sustainable Business Forum North Denmark

In one of the other new Danish regions, North Denmark, financial support for a somewhat similar network – Miljøforum Nordjylland (Sustainable Business Forum North Denmark, SBFND) – was obtained through two grants; one from the regional Growth Forum (Vækstforum), and another from the Science and Enterprise Network (ForskerKontakten), which demands strong academic involvement in the projects it supports. In that sense, this network is therefore not a P3 but rather a PPAP, and its mission is to act as a catalyst for the production and dissemination of appropriate and useful knowledge, the development of new business opportunities and the creation of a better environment (MFN, 2007). The aim of the network is to promote business development through greening – the catch phrase is 'Clean and Competitive' (MFN, 2008).

The network was established in late 2006 on the basis of a group of quality and environment managers working in the region and meeting informally from time to time. The aim was to enlarge and formalize the meetings within this group of professionals as well as including new actors in the work. It had been tried before, with the county as the leading initiator, but always failed (MFN, 2008). The reasons for these failures can only be attributed to the public bodies' (municipalities and county) resistance to collaborating with each other.

From the outside, the network could be viewed as quite similar to the networks in Southern Denmark. It is organized around a number of members from the public and private sectors in the region and supported by a secretariat. Two important differences, however, make the network unique: the strong involvement of academia (from Aalborg University, AAU), and the lesser commitment and involvement from municipal authorities. Resistance to collaborating with each other is thus still evident, even though the county has now been abolished.

In fact, the municipal authorities in the network are not pivotal and the activities are neither built around the authorities' obligations of inspecting private companies and issuing permits, nor around any dialogue between the public and the private sector that these obligations may or may not produce (MFN, 2008). Instead, the technical departments of the local government members are on the receiving end of the network's services. Any role the municipalities may have typically associated with their business development functions. This situation is also evident in the way the network's support function has been organized. This was until recently located at Væksthus Nordjylland, the Business Link Centre in the North Denmark Region.

According to the coordinator of the network, the pivotal mechanism is in fact the contact with Aalborg University. This contact includes student projects, researcher involvement in development projects and speeches and lectures at all seminars and workshops. The strong connection results in professionally and technically well-founded activities, but also stronger public–private–academic dialogue.

Besides being one of the main initiators of the network, the strong university involvement is also evident in the composition of the steering committee. Currently there are eight persons on the committee, including three faculty members from the university (one full professor and head of department, and two associate professors). The university is a full member of the network, and directs its activities to connect (theoretical) university education with practical knowledge and activity.

A brief overview of the activities in the network is described Table 3.2. These activities all have a tendency to be rather short term (at most 6 months) and based very much on experience exchange and on connecting the various actors. Any longer term interaction between, for example, a company and university faculty is therefore up to the partners themselves. In that sense, the network functions as some sort of brokerage arrangement for knowledge transfer (see, e.g., von Malmborg, 2004).

Table 3.2 Overview of activities in SBFND

Category	Activities
Cleaner Products	Support to review and policies (EMS), dialogue
(Continuing) Education	Short courses, seminars and workshops, problem-based learning (as taught at AAU)
Projects	Masters-level theses and projects; Lean, TPM; food product safety; eco-design
Information & communication	Newsletters, press releases, internet website

Recent developments have resulted in the network being relocated from its previous location at the Business Link Centre and integrated into Aalborg University. The move will be officially completed early in 2009 and shows commitment on the part of the university to these types of partnerships.

3.4 Discussion

As one of the options in the pursuit of sustainable development, the notion of P3 emerged. Academics have played a substantial role in writing background papers at WSSD, but they were later squeezed out when politicians took over:

And this remained the pattern with WSSD 2002, even though WSSD 2002 highlighted the importance of partnerships. Emphasis was, however, placed primarily on partnerships between business, government agencies and non-governmental organizations, rather than with academics (Fincham, Georg, & Nielsen, 2005).

If that is really the case, what roles might universities play in partnerships for sustainable development? Do universities have something special to offer? Are there any substantial experiences to fall back on? Much anecdotal evidence in the literature seems to suggest this contribution and role, see, for example, Gaardhøje, Hansen, & Thulstrup (2006), AAU (2001, 2002), Fincham and Korrûbel (2003), Fincham et al. (2005), Jamison and Muchie (2005), Jeppesen, Nielsen, & Fincham (2005), Hansen et al. (2005), Hansen and Lehmann (2006), and Lehmann and Fryd (2008).

Since universities, because of their third mission (engagement in the surrounding society; in this chapter exemplified by the two partnership cases presented) and the growing importance of Mode-2 (Nowotny et al., 2001) and triple-helix (Etzkowitz & Leydesdorff, 2000) knowledge production, more and more are supposed to cooperate not only with science-based high-tech firms but also with small and medium sized enterprises and other organizations (including government) engaged in low- and medium-tech activities, they are also increasingly regarded as vital drivers of regional growth and development. They are becoming crucial elements in regional innovation systems. Universities are now not only national institutions meeting the needs for science and higher education for the country as a whole. They have come to play crucial roles as knowledge hubs for specific regions including those without very much science-based production. Sometimes regional universities have been established because of strong pressure on the national government from local and regional firms and labor market organizations, branch organizations, local and regional governments.

As we have seen in the cases described in this chapter there are activities that can be described as P3 or PPAP (triple-helix) depending on the degree and character of university involvement. The two networks presented in this chapter both focus their efforts on addressing issues of sustainable development. PPAPs exist and in various ways they may constitute crucial parts of innovation systems.

In another paper (Johnson & Lehmann, 2006) we have discussed the concept of a sustainable innovation system, which was defined in the following way:

> A Sustainable Innovation System is constituted by human, natural and social elements and relationships, which interact in the production, diffusion and use of new and socially, environmentally, economically and institutionally useful knowledge that contributes to sustainable production and consumption patterns. (Johnson & Lehmann, 2006, p. 18)

To achieve sustainable development, investment is needed in four different but related types of capital; natural capital, human and intellectual capital, production capital and social capital. Human society depends on these capital stocks and without their maintenance the ability of future generations to fulfill their needs is impaired. In short, these four capitals may be described in the following ways:[3]

- *Natural capital* refers to natural resources and ecosystems. In addition to renewable and non-renewable natural resources it also includes geographical factors like climate, disease ecology and distance to the coast (for example, if a country is landlocked or not) which recent empirical research has shown to be strongly correlated with development (Sachs & Malaney, 2002).
- *Human and intellectual capital* refers to the health, education, knowledge and competence of people.
- *Production capital* is the stock of buildings, tools and machines used in the production of goods and services. This is what economists traditionally refer to as capital.
- *Social capital* is composed of the institutions which form the language, trust and networks that make continual social interaction possible.

One implication of envisaging sustainability in terms of these four capitals is that it is a process and not a state of balance or equilibrium. Even if society would

---

[3] A note on the terminology: It may be problematic to use the term 'capital' in the way we do here since it often refers to a stock which can grow or decline. Because of the diversity, incomparability and complexity of the elements in these four stocks it is often impossible to measure the size and change of them in meaningful ways. For example, social capital, which is defined as a set of rules, habits and norms, is very difficult to imagine as a stock, and how would one aggregate climate, oil and biodiversity into a single stock of natural capital? However, the use of the notion of capital has become quite common in these connections and we may think of it rather as a collection of different things than as a homogenous stock.

aim at keeping natural capital, social capital and production capital as they are, by reproducing the present situation, human and intellectual capital would change. Knowledge never rests. When time is introduced as a factor, production, consumption and social interaction implies new experiences, which result in knowledge change through learning and forgetting. Development is inevitable and sustainable development requires that the human and intellectual capital is further developed and applied to the maintenance and change of the three other capitals as well. Introduction and utilization of new or at least recombined knowledge into society is called innovation, and innovation is the only key there is to sustainability.

The two partnerships presented above are both examples of efforts to build considerable capabilities to move toward sustainability by utilizing (and creating) human and intellectual capital. They are both arenas of shared responsibility.

In one partnership, the public and private sector play leading roles and have established a governance network, but from time to time they bring in academia to create innovations and move from one level of the Greening Triangle (see Fig. 3.1) to the next (from environment to sustainability, for example).

In the other collaboration, it is the university that plays the leading role in getting the public and the private together. It is also the university that delivers administrative support providing for an even stronger function of the university as a hub for sustainable development. It has not been possible to secure reliable information as to the reasons for this move, and it is thus not known if there are strategic decisions behind the transition. One thing is certain, however, the possibility to connect the SBFND to other collaborations also located at AAU has now increased. These include, for example, the International Centre for Innovation (ICI) and the Centre for Regional Development (Center for Regional Udvikling, CRU).

As both approaches are geographically limited it makes sense to include them as aspects of special types of sustainable regional innovation systems.

3.5 Conclusions

As the two partnerships are very different in timescale (one is more than 15 years old, the other only three) and have different origins (one P3, the other PPAP), it is difficult to say whether or not they eventually will converge into a similar type of partnership and provide the same functions to sustainable regional innovation systems. These functions can be both in terms of governance, i.e., defining problems and coordination, and of an innovative nature, i.e., developing and providing new solutions to overcome problems.

However, arguing that the problems of sustainable development (as well as the solutions to these problems) are drivers of PPAPs and their innovative capabilities, we suggest the notion of *sustainable regional innovation system* to denote regional innovation systems that include significant public–private–(academic) partnerships that address issues of sustainability.

The possibilities to make development more sustainable by changing human actions and behavior are not neatly connected to the geographical/administrative level on which the problems emerge or have their impact. Governance and policies are conducted at local, regional, national and global levels and sustainable development depends on actions on all these levels. This means that regional policy-making and governance is interesting from a sustainability point of view, not because these problems are predominantly regional but rather because it is often feasible to attack them at this level. Often we find universities mediating these pressures on the companies and translating the content thereof to the members of the partnership. In some regions it may simply be possible to make political decisions and create governance structures that address sustainability issues, for example, by stimulating specific technical, organizational and institutional innovations. The literature about regional systems of innovation, as well as the cases of partnerships of the type discussed in this chapter, attests to that.

To build sustainable regional innovation systems requires the introduction of and support for sustainable development at all levels (not only the level of the home region) as a responsibility and a political goal, and to support the establishment of new governance structures like public–private–academic partnerships with a sustainability agenda.

References

AAU. (2001). *Development and research: New opportunities and threats.* Aalborg: Aalborg University.

AAU. (2002). *Development research & education: AAU strategic perspectives.* Aalborg: Aalborg University.

Asheim, B. T., & Gertler, M. S. (2005). The geography of innovation. In J. Fagerberg, D. C. Mowery, & R. R. Nelson (Eds.), *The Oxford handbook of innovation* (pp. 291–317). New York: Oxford University Press.

Bazzoli, G. J., Stein, R., Alexander, J. A., Conrad, D. A., Sofaer, S., & Shortell, S. M. (1997). Public–private collaboration in health and human service delivery: Evidence from community partnerships. *The Millbank Quarterly, 75*,(4), 533–561.

Buse, K., & Walt, G. (2000). Global public–private partnerships: Part 1 – A new development in health?. *Bulletin of the World Health Organisation, 78*(4), 549–561.

Christensen, P., Thrane, M., Jørgensen, T. H., & Lehmann, M. (2009). Sustainable development: Assessing the gap between preaching and practice at Aalborg University. *International Journal of Sustainability in Higher Education, 10*(1), 4–20.

Cooke, P. (1992). Regional innovation systems: Competitive regulation in the New Europe. *Geoforum, 23*(3), 365–382.

Cummings, A. (2005). *Against all odds. Building innovative capabilities in rural economic initiatives in El Salvador.* PhD Dissertation, Aalborg University, Denmark.

Dosi, G. (1999). Some notes on national systems of innovation and production and their implication for economic analysis. In D. Archibugi, J. Howells, & J. Michie (Eds.), *Innovation policy in a global economy* (pp. 35–48). Cambridge: Cambridge University Press.

Edquist, C. (2005). Systems of innovation: Perspectives and challenges. In J. Fagerberg, D. C. Mowery, & R. R. Nelson (Eds.), *The Oxford handbook of innovation* (pp. 181–208). New York: Oxford University Press.

Edquist, C., & Johnson, B. (1997). Institutions and organisations in systems of innovation. In C. Edquist (Ed.), *Systems of innovation: Technologies, institutions and organizations.* London: Pinter/Cassell Academic (republished in Edquist & McKelvey, 2000).

Edquist, C., & McKelvey, M. (Eds.). (2000). *Systems of innovation: Growth, competitiveness and employment.* London: Edward Elgar.

Etzkowitz, H., & Leydesdorff, L. (2000). The dynamics of innovation: From national systems and 'Mode 2' to a triple helix of university-industry-government relations. *Research Policy, 29*(2), 109–123.

Fincham, R., & Korrûbel, J. (2003). Promoting sustainable urban livelihoods and environments: The role of government, business and the public. *Proceedings environmental management & technology: A clean environment towards sustainable development, 4–6 August 2003.* Putrajaya: LUCED-I&UA.

Fincham, R., Georg, S., & Nielsen, E. H. (2005). Universities and the dilemmas of sustainable development. In R. Fincham, S. Georg, & E. H. Nielsen (Eds.), *Sustainable development and the university: New strategies for research, teaching and practice* (pp. 13–33). Howick: Brevitas.

Gaardhøje, J. J., Hansen, J. Aa., & Thulstrup, E. W. (Eds.). (2006). *Capacity building in higher education and research on a global scale.* Copenhagen: The Danish National Commission for UNESCO.

Glaeser, E. L. (2000). The future of urban research: Non-market interactions. In W. G. Gale & J. R. Pack (Eds.), *Brookings-Wharton papers on urban affairs* (pp. 101–149). Washington DC: The Brookings Institute.

Glaeser, E. L., & Scheinkman, J. A. (2001). Non-market interaction. *NBER Working Paper Series* (8053). Cambridge: National Bureau of Economic Research.

Glasbergen, P. (1998). Modern environmental agreements: A policy instrument becomes a management strategy. *Journal of Environmental Planning and Management, 41*(6), 693–709.

Glasbergen, P. (1999). Tailor-made environmental governance: On the relevance of the covenanting process. *European Environment, 9*(2), 49–58.

Hansen, J. Aa., & Lehmann, M. (2006). Agents of change: Universities as development hubs. *Journal of Cleaner Production, 14*(9–11), 820–829.

Hansen, J. Aa., Lindegaard, K., & Lehmann, M. (2005). Universities as development hubs. In R. Fincham, S. Georg, & E. H. Nielsen (Eds.), *Sustainable development and the university: New strategies for research, teaching and practice* (pp. 92–122). Howick: Brevitas.

Jamison, A., & Muchie, M. (2005). Building bridges, making spaces: On the emerging arenas of green knowledge. In R. Fincham, S. Georg, & E. H. Nielsen (Eds.), *Sustainable development and the university: New strategies for research, teaching and practice* (pp. 123–136). Howick: Brevitas.

Jensen, M. B., Johnson, B., Lorenz, E., & Lundvall, B.-Å. (2007). Forms of knowledge and modes of innovation. *Research Policy, 36*(5), 680–693.

Jeppesen, S., Nielsen, E. H., & Fincham, R. (2005). Networking and partnerships at the local level – Striving for academic relevance. In R. Fincham, S. Georg, & E. H. Nielsen (Eds.), *Sustainable development and the university: New strategies for research, teaching and practice* (pp. 276–290). Howick: Brevitas.

Johnson, B. (1992). Institutional learning. In B.-Å. Lundvall (Ed.), *National innovation systems: Towards a theory of innovation and interactive learning* (pp. 23–44). London: Pinter Publishers.

Johnson, B. (2007). Systems of innovation, the urban order and sustainable development. *Waste Management & Research, 25,* 208–213.

Johnson, B., & Lehmann, M. (2006). Sustainability and cities as systems of innovation. *DRUID Working Paper* (No. 06-17), Danish Research Unit for Industrial Dynamics.

LaFrance, J., & Lehmann, M. (2005). Corporate awakening – Why (some) corporations embrace public–private partnerships. *Business Strategy & the Environment, 14*(4), 216–229.

Lehmann, M. (2006). Government-business relationships through partnerships for sustainable development: The Green Network in Denmark. *Journal of Environmental Policy & Planning, 8*(3), 235–257.

Lehmann, M. (2008). *Conceptual developments & capacity building in environmental networks: Towards public–private–academic partnerships.* PhD Dissertation, Aalborg University, Denmark.

Lehmann, M., Christensen, P., & Larsen, J. M. (2005). Self-regulation and new institutions: The case of Green Network in Denmark. In S. Sharma & A. Aragon-Correa (Eds.), *Environmental strategy and competitive advantage* (pp. 286–308). Northampton, MA: Edward Elgar Academic Publishing.

Lehmann, M., Christensen, P., Du, X., & Thrane, M. (2008). Problem-oriented and project-based learning (POPBL) as an innovative learning strategy for sustainable development in engineering education. *European Journal of Engineering Education, 33*(3), 281–293.

Lehmann, M., & Fryd, O. (2008). Urban quality development and management: Capacity development and continued education for the sustainable city. *International Journal of Sustainability in Higher Education, 9*(1), 21–38.

Lundvall, B.-Å., Johnson, B., Andersen, E. S., & Dalum, B. (2002). National systems of production, innovation and competence building. *Research Policy, 31*(2), 213–231.

Manring, S. L. (2007). Creating and managing interorganizational learning networks to achieve sustainable ecosystem management. *Organization & Environment, 20*(3), 325–346.

MFN. (2007). *Miljøforum Nordjylland – den nyeste viden indenfor miljø.* Erhverscenter Nordjylland: Aalborg.

MFN. (2008). personal communication (June 2008) with co-ordinator Henrik Riisgaard.

Molas-Gallart, J., Salter, A., Patel, P., Scott, A., & Duran, X. (2002). *Measuring third stream activities: Final report to the Russell group of universities.* Science Policy Research Unit, University of Sussex, Brighton, UK.

Mowery, D., & Sampat, B. N. (2005). Universities in national innovation systems. In J. Fagerberg, D. C. Mowery, & R. R. Nelson (Eds.), *The Oxford handbook of innovation* (pp. 209–239). New York: Oxford University Press.

Nelson, J., & Zadek, S. (2000). *Partnership alchemy: New social partnerships in Europe.* Copenhagen: The Copenhagen Centre.

Nowotny, H., Scott, P., & Gibbons, M. (2001). *Re-thinking science – Knowledge and the public in an age of uncertainty.* Cambridge: Polity Press.

Pierre, J. (Ed.). (1998). *Partnerships in urban governance: European and American experiences.* Hampshire: Palgrave Macmillan.

Polanyi, M. (1966). *The tacit dimension.* New York: Anchor Day Books.

Roome, N. (2001). Editorial: Conceptualizing and studying the contribution of networks in environmental management and sustainable development. *Business Strategy and the Environment, 10*(2), 69–76.

Sachs, J. D., & Malaney, P. (2002). The economic and social burden of malaria. *Nature, 415*(6872), 680–685.

Schneider, A. L. (1999). Public–private partnerships in the U.S. prison system. *American Behavioural Scientist, 43*(1), 192–208.

Scott, R. W. (2001). *Institutions and organizations* (2nd ed.). Thousand Oaks, CA: Sage Publications.

Siebenhüner, B. (2005). The role of social learning on the road to sustainability. In J. N. Rosenau, E. U. Weizsäcker, & U. Petschow (Eds.), *Governance and sustainability* (86–99). Sheffield: Greenleaf Publishing.

Sjöberg, U. (1993). *Institutional relationships – The missing link in the network approach to industrial markets.* Paper presented at the 9th IMP Conference, Bath, UK.

Sorensen, O. J. (1994). Government-business relations: Towards a partnership model. *International Business Economics Working Paper Series, No. 9.* Aalborg: Aalborg University.

van Kerkhoff, L., & Lebel, L. (2006). Linking knowledge and action for sustainable development. *Annual Review of Environment & Resources, 31*, 445–477.

von Hippel, E. (1994). Sticky information and the locus of problem solving: Implications for innovation. *Management Science, 40*(4), 429–439.

von Malmborg, F. (2003). Conditions for regional public–private partnerships for sustainable development – Swedish perspectives. *European Environment, 13*(3), 133–149.

von Malmborg, F. (2004). Networking for knowledge transfer: Towards an understanding of local authority roles in regional industrial ecosystem management. *Business Strategy and the Environment, 13*(5), 334–346.

Chapter 4
Obstacles to and Facilitators of the Implementation of Small Urban Wind Turbines in the Netherlands

Linda M. Kamp

Abstract In this chapter, we combine the 'functions of innovation systems' approach and the 'socio-technical systems' approach. We first consider whether seven functions of the innovation system surrounding the technology have been fulfilled: market formation, entrepreneurial activity, knowledge creation, knowledge diffusion, mobilisation of resources, presence of advocacy coalitions, and guidance of the search. We then investigate the availability of a 'space' for this new niche technology within the incumbent energy system. We apply our framework to the development and implementation of small urban wind turbines in the Netherlands in the period 2000–2007. We show that critical functions, such as knowledge diffusion and market formation, were underdeveloped and that serious bottlenecks were present in the incumbent energy system. Based on this case study, we formulate implications for collaboration and for policy makers.

Keywords Small urban wind turbines · Functions of innovation systems · Socio-technical systems · The Netherlands

4.1 Introduction

Reducing emissions of CO_2 and other environmentally unfriendly gases is an important issue and a great challenge. For this reason, several new energy technologies (e.g. wind power, photovoltaic solar power, combined heat and power plants) have been developed in recent decades. The implementation of such new technologies often does not work out in practice, however, nor does it take place as quickly as many might hope. For example, the implementation goals for wind power and photovoltaic solar power in the Netherlands are far from being achieved.

L.M. Kamp (✉)
Delft University of Technology, Jaffalaan 5, Delft 2628 BX, The Netherlands
e-mail: l.m.kamp@tudelft.nl

J. Sarkis (eds.), *Facilitating Sustainable Innovation through Collaboration*,
DOI 10.1007/978-90-481-3159-4_4, © Springer Science+Business Media B.V. 2010

A number of broad socio-technical frameworks have been developed in recent decades to analyse the development and introduction of new technologies and for explaining why certain introduction processes have 'failed' and others have 'succeeded'. These frameworks include the socio-technical systems approach, the strategic niche management framework and the functions of innovation system approach. They have been used in several recently published analyses of the introduction of renewable energy technologies (see, e.g. Kamp, 2002; Negro, Hekkert, & Smits, 2007; Raven, 2005). In this chapter, we combine these frameworks into a single framework.

We apply this combined framework to the development and implementation of small urban wind turbine technology in the Netherlands in the period 2000–2007. Our research question is as follows:

What are obstacles to and facilitators of the Dutch small urban wind turbine innovation system?

The case of small urban wind turbine technology (UWT) was investigated using a qualitative case study methodology. We consulted literature studies, including scientific papers and reports, as well as newspaper clippings, popular magazines, manufacturers' leaflets and other grey literature. We also conducted internet investigations and held interviews with relevant actors and stakeholders.

This chapter is structured as follows. Section 4.2 contains a description of the analytical framework we used for our analysis. Section 4.3 describes the case study. Section 4.4 concludes by answering the research question and formulating implications for collaboration and policy-making.

4.2 Analytical Framework

A number of socio-technical studies on the introduction of new technologies have shown that the success of a new technology is not only determined by technical characteristics but also by the social system that develops and implements (or refuses) the new technology. Most recent socio-technical research on the introduction of renewable energy technologies uses one of the following three conceptual approaches: the functions of innovation systems approach, the socio-technical systems approach and the strategic niche management approach.

4.2.1 The Functions of Innovation Systems Approach (FIS)

In the theoretical framework of innovation systems, the social system surrounding a technology is known as a 'technology-specific innovation system' or 'TIS' (Negro et al., 2007). Carlsson and Stanckiewicz (1991) define a TIS as 'a dynamic network of agents interacting in a specific economic/industrial area under a particular institutional infrastructure and involved in the generation, diffusion and utilisation of technology'. The system consists of three main elements (Kamp, 2002; Lundvall, 1992): (1) (networks of) actors and organisations, (2) formal, normative and cognitive rules (i.e. institutions) and (3) learning processes between the actors.

Recent research using the innovation systems approach has led to the development of the notion of 'functions of innovation systems': functions that innovation systems should fulfil in order to introduce new technologies successfully (Jacobsson & Bergek, 2004; Negro et al., 2007). Several distinct sets of functions have been proposed in the recent literature. In this chapter, we use the set of functions proposed in Negro and colleagues (2007). This set will enable us to compare our case analyses at a later stage with cases on biomass technologies analysed by Negro and colleagues. The set of functions is as follows: entrepreneurial activities, knowledge development, knowledge diffusion, guidance of the search, market formation, resources mobilisation and support from advocacy coalitions. We describe each of the functions below.

4.2.1.1 Function 1: Entrepreneurial Activities

Entrepreneurs are crucial to a well functioning innovation system. Their role is to turn the potential of new knowledge, networks and market into concrete business for the new technology. They can be either new entrants to the market or incumbent companies that diversify to the new technology.

4.2.1.2 Function 2: Knowledge Development

Another very important function is the generation of knowledge or learning (Kamp, 2002; Lundvall, 1992). In earlier work (Kamp, 2002; Kamp, Smits, & Andriesse, 2004), we focused on the role of learning processes within TIS. Whereas most analyses based on the functions of innovation systems approach have focused on R&D or learning by searching, we explicitly distinguish four kinds of learning processes: learning by searching, learning by doing, learning by using and learning by interacting. Learning by searching takes place at research institutes and research departments in companies. It consists of the systematic and organised search for new knowledge, or the innovative combination of old and new knowledge. Learning by doing takes place in companies and consists of increasing production skills, resulting in an increase in the efficiency of production operations (Rosenberg, 1982). Learning by using takes place during the utilisation of the technology. Learning by using may result in knowledge about the new technology that could not have been predicted by scientific knowledge or techniques.

4.2.1.3 Function 3: Knowledge Diffusion

The fourth type of learning we distinguish is learning by interacting, or knowledge diffusion. This learning process involves the transfer of knowledge between different actors. Particularly in complex innovation processes, firms are hardly ever able to have or develop all the required knowledge and skills in-house. Successful innovation is largely dependent on close and persistent user-producer contacts. Knowledge diffusion is difficult and must occur during direct face-to-face contacts, particularly if the required knowledge is tacit and if it is difficult to formalise and communicate to a broader audience. Successful knowledge diffusion requires the fulfilment of such conditions as mutual interest in the learning process and norms of openness and disclosure, as well as proximity in the broadest sense, including geographical closeness, cognitive closeness and a common language and culture (Kamp, 2002, Kamp et al., 2004).

4.2.1.4 Function 4: Guidance of the Search

During technology development, it is impossible to explore every possible developmental path. Because resources are limited, specific paths or foci must be chosen. One example is a 'technological guidepost' (Sahal, 1981), which is a technological example that has proven to work.

We can observe this function from three angles. One angle involves the entrepreneurs and their backgrounds; another considers the guidance provided by universities and other independent research centres, and the last concerns the role of government in the form of subsidies and political pressure.

4.2.1.5 Function 5: Market Formation

In addition to entrepreneurs, a market must be present for a technology to become successful. Because it is difficult for new technologies to compete with incumbent ones, it is important to create a protected market space, or niche. Market formation is largely driven by three factors. The first factor is a demand for the least expensive and most efficient product that addresses a direct need or provides a solution to a direct problem of that same user base. The additional two driving factors are governmental subsidies and a certain image (e.g. an environmentally friendly image).

4.2.1.6 Function 6: Mobilisation of Resources

Supporting all of the activities within a TIS requires resources in the form of financial and human capital. Physical resources (e.g. materials and energy) are also needed to produce the technical objects.[1]

---

[1]Most analyses based on the functions of innovation systems do not mention these types of resources. It is nonetheless a crucial factor in the development of certain technology. The current shortage of silicon in the Japanese photovoltaic solar power innovation system is one good example.

4.2.1.7 Function 7: Support from Advocacy Coalitions

Advocacy coalitions are needed to open a space for the new technology within the incumbent regime. This is complicated by vested interests, sunk investments, regulations and routines. Advocacy coalitions are needed to create legitimacy for the new technology, counteract resistance to change and mobilise resources in the form of investments or public subsidies.

4.2.1.8 Virtuous and Vicious Cycles

These functions are obviously not independent they are interlinked, and they influence each other. For example, resource mobilisation assists knowledge development and market formation. The functions may also influence each other in a circular way, creating self-reinforcing virtuous or vicious cycles (Hekkert et al., 2007). A so-called 'motor' exists when the functions strengthen each other in a positive feedback loop, known as a virtuous cycle. The presence of motors of change within an innovation system is very important. Virtuous cycles are considered the driving forces behind well-functioning innovation systems, while vicious cycles hamper the diffusion of technology and may even lead to its collapse (Hekkert et al., 2007). It is possible to overcome vicious cycles, and it is not uncommon for virtuous and vicious cycles to alternate within a single innovation system.

4.2.1.9 Relative Importance of Functions

One question that comes to mind is whether all functions are equally important. Must all functions be present for an innovation system to function well? Are some functions crucial? How should the 'well functioning' of an innovation system be defined and measured? A number of initial ideas on the relative importance of functions have been proposed in recent literature. Based on a number of case studies, Hekkert and Negro (2008) write that the main functions appear to be market formation, entrepreneurial activities and guidance of the search. A great deal remains unknown, however, about the relationship between the importance of functions and the phase of development of the technology under scrutiny, an issue strongly emphasised by Jacobsson and Bergek (2004). They state that the importance of each function is expected to vary in time depending on development phase of the technology (Bergek et al., 2008).

4.2.2 The Socio-Technical Systems Approach

In the socio-technical systems approach, the social system around a technology is categorised into three levels (Geels, 2005): the socio-technical landscape (macro level), the socio-technical regime (meso level) and the niche (micro level). The socio-technical landscape is the exogenous environment, which usually changes slowly. It influences dynamics at the niche and regime levels, but it cannot be easily influenced by those dynamics. Examples include oil resources or the greenhouse effect. Niches are the places where new technologies emerge. In these niches, new technologies are shielded from mainstream market selection, either because they are focused on a specific part of the market or because they are protected by public subsidies (Kemp, Schot, & Hoogma, 1988). The socio-technical regime is the level of the technology (or technologies) that is currently on the market. For the energy system, the socio-technical regime would be the current power production system, which is based mainly on fossil fuels.

From the socio-technical systems approach, we derive the insight that the successful introduction of a new technology requires the existence of opportunities or openings on all three levels (Geels, 2005). Developments at all three levels must link up and reinforce each other (Verbong & Geels, 2007). While a new, potentially well-fitting technology is being developed within the niche, developments in the socio-technical landscape must work in favour of the new technology, and developments in the socio-technical regime must create an opening for the new technology to enter the market. The latter condition is particularly difficult to fulfil, as socio-technical regimes are characterised by path dependence and lock-in into existing technologies. This is a result of sunk investments, the vested interests of actors in the regime, and the current regulations and cognitive routines of the actors that support the incumbent technologies within the regime (Jacobsson & Bergek, 2004; Unruh, 2002).

4.2.3 Strategic Niche Management (SNM)

Kemp et al. (1998) define strategic niche management as 'the creation, development and controlled phase-out of protected spaces for the development and use of promising technologies by means of experimentation, with the aim of (1) learning about the desirability of the new technology and (2) enhancing the further development and the rate of application of the new technology'. Three important processes are considered crucial within the framework (Raven, 2005): the voicing and shaping of expectations, network formation and learning processes. As described by Hoogma (2000), important aspects of network formation include network composition, network alignment and the presence of macro actors. Macro actors are those that play a leading role within a niche and increase alignment.

Figure 4.1 shows a schematic overview of our combined analytical framework. The technology that we analyse is located within a technological niche at the lower level of the figure. The middle level shows the socio-technical regime, which consists of incumbent technologies. The upper layer shows the landscape, which consists of slow developments at the national or global level that cannot be influenced easily by actors within the niche.

The socio-technical systems approach regards all three layers in the analysis. The functions of innovation systems (FIS) approach and the strategic niche management (SNM) approach regard processes within the technological niche. This is

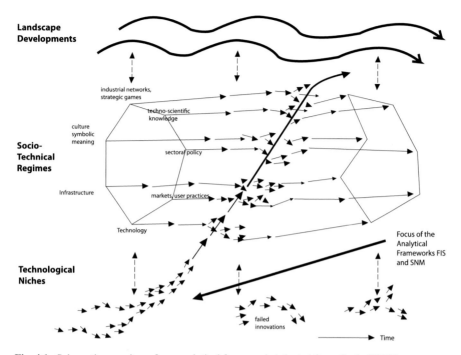

Fig. 4.1 Schematic overview of our analytical framework (adapted from Geels (2005))

represented in Fig. 4.1 by the thick black arrow. By using these approaches in combination, we try to present a complete picture of obstacles to and facilitators of the technology within the niche.

4.3 Small Urban Wind Turbines (UWTs) in the Netherlands

In this section, we analyse our case study.[2,3] We first consider each of the innovation system functions. Second, we analyse the extent to which developments at the niche level (the innovation system), the socio-technical regime level (the incumbent power regime based on fossil fuels) and the socio-technical landscape level linked up and reinforced each other. We then reflect on the usefulness of the analytical framework for our analysis.

---

[2]Because of space constraints in this chapter, we describe our analysis in a very concise manner. For a more extensive case description and analysis, see Kamp and Jerotijevic (forthcoming) or contact the author.

[3]The majority of the case study material was collected by Milutin Jerotijevic, an MSc student in Civil Engineering at TU Delft.

4.3.1 Functions of Innovation Systems

4.3.1.1 Function 1: Entrepreneurial Activities

The Dutch UWT market started at the World Expo in 2000 where Tulipo, a small turbine from Lagerweij (now known as WES), was exhibited on the roof of the Dutch pavilion (Wineur, 2007). In 2007, 14 UWT suppliers were active in the Netherlands. Eight of the suppliers manufactured their own products, while the others were importing UWTs. At that time, 56 UWTs were installed across the country (Wineur, 2006b). Of particular importance is that 37 out of the 56 installed turbines were manufactured by two suppliers: Turby with 20 and Fortis Wind Energy with 17.

Fortis Wind Energy focuses most of its efforts on remote locations and developing countries (Klimbie, 2007). Their main products have been developed specifically for that market. If we consider that this company claims over 6000 turbines installed worldwide (Klimbie, 2007), the number sold for the built environment in the Netherlands is not even 1%. Another significant manufacturer in the Netherlands is WES. During our interview, it became clear that WES is not interested in the market concerning the built environment, and it openly refers to urban wind as a myth. Turby is currently the only major manufacturer whose efforts are strictly focused on the built environment. Its product is still in the pilot project stage, however, and it is not open to wide commercial use (Sidler, 2007). There are difficulties with electronics, vibrations and other issues, and the company is still improving according to the feedback of early customers.

4.3.1.2 Function 2: Knowledge Development

The majority of knowledge development in the Netherlands concerning wind energy takes place at TU Delft, the R&D departments of suppliers and at the research institutes TNO[4] and ECN. [5] Knowledge development at TU Delft takes place primarily in the form of learning by searching. The research focuses on wind potential in the built environment, wind flows around buildings, efficiency and the optimisation of turbines through aero-elastic modelling (Ummels, 2007). Knowledge development activities have also concerned social and political factors.

A number of universities of applied sciences in the Netherlands are also involved in knowledge development. One good example is The Hague University of Applied Sciences, which developed a prototype device to measure the wind speed on rooftops (Vries de & van der Horst, 2006). Another example is Windesheim University of Applied Sciences, which performed a study concerning the permit process in 10 locations focusing on such matters as identifying the greatest problems

---

[4]TNO is a Dutch research institute. The abbreviation stands for Applied Scientific Research (Toegepast Natuurwetenschappelijk Onderzoek).

[5]ECN is a Dutch research institute. The abbreviation stands for Energy research Centre of the Netherlands (Energie onderzoekscentrum Nederland).

and finding ways to expedite the process. Research into UWTs has also been conducted at TU Eindhoven. In addition to universities and schools in the Netherlands, research has also been carried out at ECN, TNO and other research institutes.

Within the R&D departments of suppliers and manufacturers, learning takes place in all three forms. It is interesting to note that learning by doing does not actually take place with the manufacturers. All of their parts are manufactured individually at different locations, where the majority of this type of learning takes place (Klimbie, 2007; Kloesmeijer, 2007; Sidler, 2007). The turbine manufacturers gain this type of specific knowledge only through interaction with manufacturers of system parts. Blades, control panels and inverters are examples of this type of parts. These relationships therefore have considerable influence on the entire growth process of UWT technology; they will be discussed in the next section.

Although learning by searching also varies from manufacturer to manufacturer, two clusters can be distinguished. One cluster includes manufacturers that invested heavily in learning by searching in the recent past, and the other cluster is composed of manufacturers focused on improving the product developed in the more distant past and proven to work. Turby and Home Energy invested considerable effort into the design of their new products (Sidler, 2007). In contrast, Fortis Wind Energy and WES invest no effort in learning by searching when developing a new turbine model specially developed for the built environment (Klimbie, 2007; Kloesmeijer, 2007). Nonetheless, they are extensively involved in research into new inverters (Klimbie, 2007; Kloesmeijer, 2007). It is important to note that research subsidies were listed as a positive factor in the search for more efficient solutions. Finally, learning by using is of great importance at the stage of early implementation and pilot projects. All three major manufacturers maintain relationships with their current customers and are constantly receiving feedback on the performance of their turbines (Klimbie, 2007; Kloesmeijer, 2007; Sidler, 2007).

Three pilot projects are currently in progress in the province of Groningen, the province of Zeeland and in The Hague. The Groningen project is a good example of the difficulties involved in the integration of this technology. This project suffered many setbacks due to non-technical issues, such as obtaining building permits. Much more can still be completed in order to speed up the process, and new solutions should be tested with each new pilot project. These technical and non-technical issues can sometimes form a vicious circle. Building permits are hard to obtain because there is not enough data on the product concerning safety, noise, vibrations and efficiency. At the same time, however, it is hard to accumulate this data without active learning by using a process that requires first obtaining permits.

Overall, 11 manufacturers are participating in the pilot project in Zeeland. The objective of this project is to compare the yield and noise production of 11 different turbines. The required permits were obtained in November 2006, and the deployed turbines are currently being monitored. Before this project, all of the manufacturers were asked to make predictions for expected energy outputs, which would subsequently be compared to the actual readings. Of all the products for the built environment, only the Tulipo was certified by the American UL certification board, but this certification expired in 2006 (Kloesmeijer, 2007). The rest of the turbines lack any type of certification. The results of this project will serve as a strong base for the possible creation of a certification board in the Netherlands. The feedback information obtained from this project will be used to locate problems and address them transparently, thus providing a backbone for a successful process of learning by using.

Finally, we mention the project in The Hague. This project is coordinated by the Renewable Energy Company (RenCom) and is still in its early phase. The city reserved 200,000 euros for the project in 2007 for this purpose (Wineur, 2007). Unlike the other two projects, the project in The Hague involves a more densely populated built environment; it can therefore be of great importance for future development in such areas.

4.3.1.3 Function 3: Knowledge Diffusion

We consider this function the most delicate. Without well functioning knowledge diffusion within new technological development, all of the learning discussed in previous section happens at a much slower rate and, in some cases, does not happen at all. Nonetheless, the importance of knowledge diffusion has often been overlooked. Each interview in this study, however, revealed evidence of an attitude of 'I will never work with them again'. Especially in the R&D phase, manufacturers found it difficult to find a favourable common ground for collaborating with the research institutes of TNO and ECN (Sidler, 2007). While the manufacturers often sought hands-on solutions for eminent problems, the research institutes tended to focus more on complete redesigns.

Knowledge diffusion between the R&D departments of different turbine manufacturers could prove quite useful during the development stage. Nonetheless, this is one of the most problematic aspects of knowledge diffusion, as it directly interferes with the notion of competition. It is therefore not surprising that Fortis Wind Energy, Turby and WES do not participate in any form of knowledge sharing (Klimbie, 2007; Kloesmeijer, 2007; Sidler, 2007). There was, however, evidence of a positive attitude of cooperation with regard to a major technical problem that could be solved: the development of inverters to transform the DC current produced by the turbines into AC current to be fed to the grid. Both Turby and Fortis are willing to join forces in order to optimise this process (Klimbie, 2007; Sidler, 2007).

The cooperation on which most turbine manufacturers rely during and after the development phase is not with other turbine manufacturers, but with manufacturers of smaller parts that are necessary for the operation of the entire turbine. For example, WES attributes their success to their good relationship with manufacturers, such as Beyers, the Swedish control panel producer (Kloesmeijer, 2007).

Moving beyond the development stage towards early implementation, different types of relationships become more important, most notably between the manufacturers and customers. As mentioned in the previous section, a number of ongoing projects exist for this purpose, and manufacturers appear to be highly aware of the importance of feedback information (Cace, 2007). Such information is especially important for new, untested products on the market. This learning process is not limited to technical difficulties; it also involves such issues as how to overcome these difficulties and how to create a more efficient administrative procedure.

The diffusion of knowledge between manufacturers and consulting firms is also crucial in both pilot and later projects. Largely through the efforts of RenCom, a committee was formed in October 2007. This committee meets every 6 weeks to discuss wide variety of issues (Cace, 2007). Their current agenda comprises four points: certification, monitoring, lobbying and informing the market/user. The success of this committee could be one of the factors that will ultimately determine the success of the overall technology, at least in the short term.

4.3.1.4 Function 4: Guidance of the Search

We can observe this function from three angles. One angle involves the entrepreneurs and their backgrounds. Another angle concerns the guidance given by universities and other independent research centres. The third angle has to do with the role of government in the form of subsidies and political pressure.

From the beginning, Turby has maintained a policy of developing a turbine for the built environment, and the guidance of search came from the Darrius turbine, an early model of Vertical Axis Wind Turbines. In contrast, Fortis Wind Energy is based on the company's expertise in the area of supplying off-grid customers in the developing world with Horizontal Axis Wind Turbines (Klimbie, 2007). This project emphasised simple solutions that required no maintenance, as the customers were usually located in remote locations (Klimbie, 2007). Very little effort has been put into optimising their product for the built environment.

The largest share of university research is currently performed at TU Delft. According to Fortis, Turby and WES, TU Delft is providing considerable guidance in the form of explaining aerodynamics in the built environment and their possible integration into buildings (Klimbie, 2007; Kloesmeijer, 2007; Sidler, 2007). The strong base of technical knowledge and reputation that has been developing at TU Delft is a very strong facilitator of UWT technology. In this case, however, guidance is not only limited to technical support services or to TU Delft. Increasing numbers of educational programmes are emphasising the search for ways to manage the integration of new technology.

Guidance by the government in the form of market and research subsidies has also played a role. We discuss this in more detail in the next section.

4.3.1.5 Function 5: Market Formation

Market formation is largely driven by three factors. The first factor is a demand for the cheapest and most efficient product that addresses a direct need or provides a solution to a direct problem. In the context of the built environment in the Netherlands, it is difficult to define direct needs that UWT technology can address for the user base.

In the Netherlands, the provision of electricity is extremely reliable. On average, electrical outages equal 18 min a year (Ummels, 2007). The average cost of electricity is 20 euro cents/kWh (Cace, 2007). Current UWTs cannot compete with this price. Due to this lack of financial competitiveness for fulfilling a direct need of the customers, the additional two driving factors are introduced in the form of governmental subsidies and eco-labels.

Three forms of market subsidies are available for UWTs: S.D.E., E.I.A. and M.E.P.[6] (Cace, 2007; Masselink, 2007). All three of these programmes involve general subsidies for renewable energy technologies. In addition, a feed-in tariff of 8.8 euro cents/kWh is available. Dutch subsidy regulations, however, are very unreliable. In the second half of 2007, many subsidy schemes were discontinued because of budget depletion. More recently, the Dutch Minister of Economic Affairs stated that she did not see the advantage of SWTs and that no special policy would be made for them. Fortunately, UWTs currently fall under the more general renewable energy subsidies mentioned above, as they specify no lower limit for power generated. Nonetheless, the long-term availability of subsidies for UWTs is highly uncertain.

The demand for UWTs is largely dependent upon their environmentally friendly image. Turby counts on the environmental conscience and innovative nature of future turbine owners. In addition to the turbines themselves, Turby also sells fear of disasters 'because we will have them unless we act' (Sidler, 2007). Unlike Fortis Wind Energy, whose efforts are directed towards reaching a more competitive price for their product against the current electricity regime, Turby relies on a rapid increase in the price of fossil fuel to make their product financially desirable.

The market niche is created at the level of local governments, large corporations, large institutes and other large actors associated with environmental policies (Klimbie, 2007; Kloesmeijer, 2007; Sidler, 2007). The financial stability of large corporations allows them to invest in technologies that have yet to be proven. By investing in green technology, a corporation can enhance its image among its customer base, thereby achieving financial benefits that exceed the original investment. Similar processes apply to municipalities and their future voters, schools and their future students and other such entities. Wind Energy Solution refuses to sell their product (Tulipo) to private individuals. This policy is intended to prevent future complications, as private individuals are more concerned with the efficiency of the turbine, which WES believes cannot be accurately predicted in the built environment (Kloesmeijer, 2007). The strategy of Fortis Wind Energy is to rely on its relationship with local municipalities that are environmentally oriented. Once a municipality has expressed interest in promoting its environmental policy, Fortis Wind Energy offers its services for studying sites that are proposed by the municipality, along with the subsequent implementation (Klimbie, 2007).

---

[6]S.D.E. – Stimulation of Sustainable Energy Production E.I.A. – Energy Investment Deduction for profit-making organisations (Wineur, 2006a) M.E.P. – Electricity Generation Environmental Quality: applies to the total energy generated by a renewable energy installation (Wineur, 2006a)

In addition to the technical immaturity of UWT technology, a number of other obstacles also limit market formation. The lack of an official certification procedure for manufacturers and their products was among the main key words emerging during each interview, as was the long and arduous process of obtaining building permits. These two problems can be attributed to the fact that UWT technology involves too many actors that can impede the process. Because there are no official certificates for such matters as structural safety, noise levels, vibrations or efficiency, each building permit requires separate analyses for each manufacturer and each individual location (Masselink, 2007). The high potential for complaints from neighbourhood residents regarding the placement of UWTs can cause considerable delays in the entire implementation process. The current niche therefore includes the less populated built environment as well as large customers. In addition to reducing the potential number of complaints and safety hazards, less-populated areas have lower level of terrain roughness and better wind predictability.

4.3.1.6 Function 6: Mobilisation of Resources

Resources include both financial and human capital. Sources of financial capital include the government, banks or venture capitalists, and energy companies. As concluded above, market subsidies, feed-in tariffs and research direct subsidies to entrepreneurs are currently available in the Netherlands. Turby received grants at various stages of product development (Sidler, 2007). Due to a change in the national government attitude concerning these subsidies, however, grants are now awarded through a tendering process, which lowers the chance of success (Sidler, 2007).

We must also consider local governments. Although local governments cannot provide feed-in tariffs, their resources can be very useful in both market and research areas. Pilot projects that are partially sponsored by municipalities serve as both a learning process and as market promotion. The above-mentioned projects in Zeeland, Groningen and The Hague are examples of this type of financial support. Such projects would also be impossible without the involvement of energy utilities. Delta Energy is a co-sponsor of the Zeeland project. Other important actors in financial resource mobilisation include Shell and other large corporations (Cace, 2007). In addition to its involvement in the pilot projects, Shell has provided funding for research. Collaboration with Turby is one example of Shell's involvement in knowledge development.

We conclude this section by briefly considering human capital. The Netherlands has a strong knowledge base concerning aerodynamics. None of the manufacturers mentioned a lack of human capital (in terms of either quality or quantity) as a major obstacle.

4.3.1.7 Function 7: Support from Advocacy Coalitions

To address the final function, we consider three possible ways in which lobbying takes place with regard to renewable sources of energy. The first obviously involves the activities that the industry uses to promote and lobby their products. The second aspect involves promotion through political parties whose ideology is consistent with the use of renewable energy. Environmental organisations and other NGOs form a final source of advocacy support.

One positive development took place in late 2007, with the formation of a committee involving company representatives, RenCom, university professors and other important players in a clear agenda. This agenda comprises the following points: creating a system of manufacturing certification in the short-term that will lead to the development of long-term certification standards and the formation of a certification commission; lobbying towards the government; informing the potential market about the products (Cace, 2007). This committee currently meets once every 6 weeks, and all the results are distributed to a wider network in a summary created by RenCom.

The only political party that is currently active in lobbying for UWT technology in the Netherlands is the green party (GreenLeft). The most recent push for further governmental involvement came at the request of the green party faction in Parliament in October 2007. As discussed in the previous section, this request was met with a declaration of lack of interest by the Minister of Economic Affairs.

Finally, we were unable to find any evidence of involvement by environmental organisations during the research for this chapter.

4.3.2 Socio-technical Systems: Landscape Regime – Niche

We now use the socio-technical systems approach to analyse factors external to the wind-power innovation system. These factors include the technological and technological-system characteristics of the innovation and network aspects, such as the presence of macro actors.

4.3.2.1 Fit Within the Landscape Regime (Including Technical Regime Aspects)

Landscape developments continue to be favourable for the development of wind power, including small urban wind turbines. The oil crises of the 1970s heightened the need to develop power production technologies that could make the countries more self-sufficient. Because the Netherlands owns a large natural gas field, however, the need to develop such new technologies is less intense than it is in some other countries. Environmental concerns (e.g. 'acid rain' in the 1980s and the greenhouse effect in the 1990s) provided continued support for the legitimacy of efforts to develop renewable power production technologies.

Developments at the regime level are always more complex. Since the 1990s, the electricity sector in the Netherlands has developed from a reliable, stable and static system, in which the structure of power was clearly centred on the large energy producing companies, into a still reliable but more unstable and dynamic system, in which the structure of power is more spread among the stakeholders, with the national government playing an increasing role. Furthermore, a free market was developed in the European Union. These developments were in addition to the decentralisation of power production units and the increasing room for new electricity-producing technologies. The technical regime aspects in this case involve grid connection issues.

Another major obstacle that is related to the technology's fit within the regime involves the presence of a certification or verification system for the manufacturers. No such system exists for small UWTs. The interviewees mentioned this as a serious obstacle. Without a certification process for manufacturers, it becomes difficult to obtain insurance permits and building owners become wary of the potential hazards of the mounted turbine. Manufacturer certification influences safety standards, as well as the quality of the turbine in terms of efficiency. The lack of an official certification procedure for manufacturers and their products was among the main key words emerging from the interviews, as was the long and arduous process of obtaining building permits. Because there are no official certificates for structural safety, noise levels, vibrations, efficiency and similar matters, it is necessary to conduct separate analyses for each manufacturer and each individual location (Masselink, 2007). The number of potential complaints from residents of areas in which UWTs are to be placed is another source of serious delay in the entire implementation process. The current niche thus includes both large customers and the less populated built environment.

4.3.2.2 Technical Characteristics

Wind-turbine technology has proven very difficult to develop. The urban environment is quite difficult as well. The built environment presents a challenge from a technical point of view, particularly because of its fluid nature. The difficulty of assessing wind behaviour around buildings is complicated by the fact that a single new building in a neighbourhood can change all the patterns of prior wind behaviour (Plumb, 2007). In addition to the unpredictability of wind speeds, however, the turbulent nature of the wind is a source of problems as well. Examples include blade fatigue and reduced energy efficiency (Klimbie, 2007; Sidler, 2007).

4.3.2.3 Network Aspects – Network Structure, Alignment, Expectations and Macro Actors

An interesting concept from the strategic niche-management approach is that of macro actors – actors that play a leading role within a technological niche. In this case, Mrs. Cace, owner of RenCom, clearly appears to be a macro actor. Mrs. Cace plays a major role in network alignment, thereby facilitating knowledge development and knowledge diffusion, in addition to lobbying for market stimulation.

4.4 Conclusion and Discussion

4.4.1 Obstacles and Facilitators

In conclusion, using our combined analytical framework we have identified the following aspects as the most important facilitators:

- A 'green image' and openings in the energy regime (liberalised electricity market): These factors combine to facilitate the market for UWT technology.
- The macro actor RenCom, which plays a major role in improving the network of actors
- Network alignment: mainly as a result of efforts of the macro actor RenCom
- Knowledge diffusion in pilot projects: in which entrepreneurs, municipalities and, in many cases, RenCom work together and learn through interaction.

The most important obstacles that we identified are as follows:

- A lack of a common guidance of the search, because the manufacturers are developing fundamentally different turbine types
- Little resource mobilisation by policy makers
- No certainty regarding subsidies
- Very little knowledge diffusion between entrepreneurs and between entrepreneurs and research institutes
- Technical aspects: The technology is difficult, as is the environment in which it must function.
- The lack of a certification procedure
- A slow permit process

4.4.2 Virtuous and Vicious Cycles

As mentioned in Section 4.2, an important condition for successful development and implementation of a technology within its socio-economic context is the presence of virtuous cycles that are self-reinforcing and therefore lead to continuous development. Vicious cycles, in contrast, delay growth or lead to the failure of the innovation system.

The case of small UWTs in the Netherlands in the period 2000–2007 provides evidence of two main vicious cycles:

1. Lack of certainty regarding market subsidies => small market => few entrepreneurs and few turbine owners => little support from advocacy coalitions => lack of certainty regarding market subsidies
2. Weak knowledge exchange with turbine owners => little technological improvement => Small market => few owners => weak knowledge exchange with turbine owners The case is also characterised by the following virtuous cycle:

Market formation abroad => entrepreneurial activities in the Netherlands => creation of working technical objects => market formation abroad

Although this cycle does lead to entrepreneurial activities in the Netherlands, it does not lead to the implementation of many small UWTs in the Netherlands.

4.4.3 Implications for Collaboration

This chapter contains numerous examples of the importance of collaboration in the case of small UWTs. Research institutes guide entrepreneurs in their R&D processes. Knowledge diffusion between various actors requires good collaboration. Within the value chain, collaboration with suppliers appears to be important. Finally, establishing advocacy coalitions in order to make an opening within the incumbent energy regime and lobby for policy measures also requires collaboration.

4.4.4 Implications for Policy-Making

What can we learn from this research for policy-making? First, policy makers should consider factors that appear to be lacking in the innovation system and which vicious cycles exist for any specific technology. Subsequent policy measures should be targeted at triggering particular factors in order to transform vicious cycles into virtuous cycles. For the case of small UWTs in the Netherlands, we have seen that uncertainty regarding market subsidies and weak knowledge exchange with turbine owners is an important factor in the vicious cycles within the innovation system. These factors should be addressed with policy measures.

Second, the context around the innovation system (in this case, regime and landscape) should provide an opening for the new technology. This means that policy measures should also focus on removing barriers in the context of the innovation system. The slow permit process is one example from this case.

References

Carlsson, B., & Stanckiewicz, R. (1991). On the nature, function and composition of technological systems. *Journal of Evolutionary Economics, 1*, 93–118.

Geels, F. W. (2005). *Technological transitions and system innovations: A co-evolutionary and sociotechnical analysis.* Cheltenham: Edward Elgar.

Hekkert, M. P., & Negro, S. O. (2008). Functions of innovation systems as a framework to understand sustainable technological change: Empirical evidence for earlier claims. *Technological Forecasting and Social Change, 76*(4), 584–594.

Hekkert, M. P., Suurs, R. A. A., Negro, S. O., Kuhlmann, S., & Smits, R. E. H. M. (2007). Functions of innovation systems: A new approach for analyzing technological change. *Technological Forecasting and Social Change, 74*(4), 413–432.

Hoogma, R. (2000) *Exploring technological niches*. Enschede: Academic Thesis Twente University.

Jacobsson, S., & Bergek, A. (2004). Transforming the energy sector: The evolution of technological systems in renewable energy technology. *Industrial and Corporate Change, 13*(5), 815–849.

Kamp, L. M. (2002). *Learning in wind turbine development – A comparison between the Netherlands and Denmark*. Utrecht: Academic Thesis.

Kamp, L. M., & Jerotijevic, M. (forthcoming). A functions of innovation systems analysis of the development and implementation of small urban wind turbines in the Netherlands. *Technological forecasting and social change*, submitted for publication.

Kamp, L. M., Smits, R. E. H. M., & Andriesse, C. D. (2004). Notions on learning applied to wind turbine development in the Netherlands and Denmark. *Energy Policy, 32* (14), 1625–1637.

Kemp, R., Schot, J., & Hoogma, R. (1998). Regime shifts to sustainability through processes of niche formation: The approach of strategic niche management. *Technology Analysis and Strategic Management, 10* (2), 175–195.

Lundvall, B. A. (1992). *National systems of innovation – Towards a theory of innovation and interactive learning*. London: Pinter Publishers.

Negro, S. O., Hekkert, M. P., & Smits, R. E. H. M. (2007). Explaining the failure of the Dutch innovation system for biomass digestion – A functional analysis. *Energy Policy, 35* (2), 925–938.

Raven, R. P. J. M. (2005). *Strategic niche management for biomass*. Eindhoven: Technische Universiteit Eindhoven.

Rosenberg, N. (1982). *Inside the black box: Technology and economics*. Cambridge: Cambridge University Press.

Sahal, D. (1981). *Patterns of innovation*. Reading, MA: Addison Wesley.

Unruh, G. C. (2002). Escaping carbon lock-in. *Energy Policy, 30* (4), 317–325.

Verbong, G., & Geels, F. (2007). Lessons from a socio-technical, multi-level analysis of the Dutch electricity system (1960–2004). *Energy Policy, 35* (2), 1025–1037.

Vries de, A., & Horst van der, M. (2006). *Wind energy: Integration in structures*. Eindhoven: University of Technology.

Wineur. (2006a). *Wind Energy Integration in the Urban Environment, Techno-Economic Report*.

Wineur. (2006b). *Wind Energy Integration in the Urban Environment: Administrative and Planning Issues, Netherlands Country Report*.

Wineur. (2007). *Guidelines for small wind turbine in the built environment*.

Interviews

– Cace, J. (2007), owner, Renewable Energy Company
– Klimbie, B. (2007), sales manager, Fortis Wind Energy
– Kloesmeijer, M. (2007), sales manager, Wind Energy Solutions
– Masselink, P. (2007), consultant, Senternovem
– Plumb, H. (2007), Assistant Professor, Architecture Faculty TU Delft
– Sidler, D. (2007), owner, Turby B.V
– Ummels, B. C. (2007), PhD, Electrical Engineering Faculty TU Delft

Chapter 5
Regional Sustainability, Innovation and Welfare Through an Adaptive Process Model

Kjell-Erik Bugge, Bill O'Gorman, Ian Hill, and Friederike Welter

Abstract Over the last 10 years or so the EU has supported many initiatives focused on enhancing regional competitiveness, regional innovation, and regional sustainability. Whilst a plethora of initiatives has been developed and presented, ongoing sustainability of regional innovation processes and regional innovation clusters still eludes us. A proposed solution is the Adaptive Model for Creating a RTD (Research and Technology Development) Investment Policy for Regions in Emerging and Developed Economies (CRIPREDE), which was developed as part of an EU FP6,[1] Regions of Knowledge 2 co-funded project. The Adaptive Model was co-developed, and tested, in a highly interactive process, involving stakeholders and research organisations in six very different (political, cultural, economic) regions across the EU.

The Adaptive Model's success is measured, in part, by the sustainable action plans that have been implemented in each of the regions involved in the project. Its development and success has been achieved through the underlying principles of the Triple-P, Triple Helix, and Entrepreneurial Imperative models. An overriding principle of the whole process is that the regional stakeholders are the drivers and owners of the regional developmental process and the implementation of the regional action plans derived from the Adaptive Model.

Keywords Triple P · Triple helix · Regional competitiveness · Regional innovation · Adaptive model

K.-E. Bugge (✉)
Saxion University of Applied Sciences, P.O. Box 501, Deventer, 7400 AM, The Netherlands
e-mail: k.e.bugge@saxion.nl

[1]FP6 is the sixth European Union (EU) Framework Programme for Research and Technology Development (RTD). Based on the Treaty establishing the EU, the Framework Programme has to serve two main strategic objectives: Strengthening the scientific and technological bases of industry and encourage its international competitiveness while promoting research activities in support of other EU policies (CORDIS FP6, 2009).

J. Sarkis (eds.), *Facilitating Sustainable Innovation through Collaboration*, 77
DOI 10.1007/978-90-481-3159-4_5, © Springer Science+Business Media B.V. 2010

5.1 Introduction

Most economies support sustainable development, a development that '[...] meets
the needs of the present without compromising the ability of future generations to
meet their own needs' (UN-WCED, 1987, p. 24). There also seems to be general
agreement that sustainability implies that some kind of balance is needed between
objectives concerning environmental protection, economic growth and social equity
(UN-WCED, 1987): a balance often referred to as the 'Triple-P' (People – Planet –
Profit) principle (Elkington, 1997).

However, the real challenge is encountered when attempts are made to opera-
tionalise sustainability according to such guidelines or principles. Finding 'ideal'
solutions would imply assessing all possible options at all scales and involved
actors as a function of time. Regardless of whether such an approach is possible,
it certainly is impractical. Therefore choices have to be made concerning scope and
focus. Each individual choice introduces its own specific dilemmas (Bugge, 2003).
Although there is a rapidly growing body of knowledge on approaches to address
sustainability, there is still a need to find well-working ways to support such difficult
development and decision-making processes. The core challenge is a process innovation aimed at 'bridging the gap' between the noble, but often abstract, principles
of sustainability and the day-to-day practice of local and regional development.

Much attention has been given to developing improved strategies at local, municipal or neighbourhood level. However, the impact of applying a wider regional
perspective should not be underestimated. Therefore the Triple-P effects need to
be addressed on this scale. This again introduces the specific challenge of developing innovative mechanisms that support planning and decision-making in highly
complex multi-actor settings. Such settings not only include local authorities, but
also actors such as industry and universities.

This chapter presents a new process approach to innovation and sustainability on
a regional scale that attempts to fully take into consideration the challenge of 'merging' a variety of, often partly conflicting, interests and ideas in a multi-actor setting.
Besides applying relevant theory, it specifically draws on the experiences and results
of a 2-year FP6[2] co-funded research project called 'Creating a RTD Investment
Policy for Regions in Emerging and Developed Economies' (CRIPREDE, 2007a).
The chapter demonstrates how sustainable regional development can be effectively
stimulated and facilitated through an integral decision-support approach to 'hard'
content and 'soft' collaboration issues. This approach has been integrated into, and
is presented as, the CRIPREDE Adaptive Model. The Adaptive Model has been
applied as a support tool in the development of action plans for RTD (we use the
terms RTD and innovation interchangeably) in six regions across Europe.

Besides its value for practitioners, this chapter aims to contribute to the body of
knowledge within the fields of regional development, innovation, and sustainability. In particular, the focus is on improving the understanding of interdependency

---

[2]See Footnote 1.

between the processes of complex interactive planning and regional sustainable innovation.

The conceptual review is introduced in the next section of this chapter. The following section describes the methodology applied for developing the Adaptive model and Regional Actions Plans. The fourth section focuses on the tangible as well as less tangible outcomes. Finally, the value of the results for theory and practice will be discussed, leading to some preliminary conclusions and suggestions for further research.

5.2 Conceptual Review

Research has identified a variety of factors influencing RTD in a regional context. Overall, one can identify learning processes and their outcomes, namely knowledge, networks, networking and the role of key actors and spatial embeddedness as main factors influencing regional RTD.

In a R&D context, knowledge is an input needed for regional RTD, while learning refers to the process underlying the transfer of tacit and non-codified knowledge into explicit and codified knowledge. Here, research has identified a technical culture as one element needed for a favourable RTD environment (Malecki, 1997). Where regions facilitate the exchange of ideas, are open for new experiences and foster networking between actors across different levels, learning in the region is supported. Such collective learning is said to be closely linked to proximity, as it is based on conversations and interactions among stakeholders within a particular context, which has led some authors to introduce the concept of the 'learning region' as a region where external knowledge flows are effectively disseminated and integrated into a region's internal systems of information diffusion (Morgan, 1997). As Lawson pointed out the discussion around collective learning is 'an attempt to trace out the mechanisms by which proximity influences innovative behaviour' (Lawson, 1997, p. 21).

This draws attention to networks and the social embeddedness of business relations. The significance of networks for the economy of regions has long been recognised in the literature on regional and local development (see, for instance, Amin & Thrift, 1994). Trust is the 'lubricant' without which such network activities at regional level would not be possible (Anderson & Jack, 2002); within a region, individuals could build up reputations of trustworthiness, which is important information for other regional actors if those 'trusted' persons participate in a newly emerging network. In this regard, some research emphasises the role(s) network actors play in and for network emergence and its further development. For example, applying the concept of innovation promoters, Koch, Kautonen, and Grünhagen (2006) showed that actors within networks often fulfil several promoter roles, mainly acting as process and relationship promoters. The works on the creative milieu suggest that so-called high communicators play an important role for network development, as they transmit information, speed up decision-making,

and foster inter-organisational linkages (Fromhold-Eisebith, 1995). These key individuals contribute to the development of 'institutional thickness' by bringing in
local knowledge and the ability to access and link local capacity at different levels (Malecki, 1997). Thus, key actors help with RTD insofar as they enhance or
build local capacity.

The last factor concerns spatial proximity and its wider influence on regional
RTD. It plays an important role in creating competitive advantages of both firms and
regions (Lechner & Dowling, 2003; Liao & Welsch, 2005; Maskell & Malmberg,
1999; Schamp & Lo, 2003). However, there also exists a 'dark' side of embeddedness, as there is a trade-off between strong networking ties within a region,
over-embeddedness and the danger of being locked-in in networks which in turn
might stifle economic performance (Welter & Kolb, 2006). Overly strong forms of
interpersonal trust might result in closed networks and inward-looking behaviour
both on an individual and regional level. Consequently, entrenchment may result,
and the performance of regions can be impeded.

In summarising previous research (see Table 5.1, and Welter & Kolb, 2006, for
more details), we recognised the importance of relating 'hard' and 'soft' input factors required for RTD to specific levels, resources, and, in particular, processes.
In this way, we attempt to capture elements of what makes regions successful in
growing and developing rather than a whole strategy for creating good practices
within regional RTD. We also recognise a need to simultaneously focus on factors and processes, as it is the interplay between both that can foster RTD. Finally,
we relate successful regional RTD to critical factors regarding the 'danger' of
over-embeddedness, and undesired lock-in effects.

We recognise overlaps, but also a need to conceptually distinguish between
three main categories of input factors (Table 5.1) influencing RTD namely, (1) general conditions and resources, (2) institutional infrastructure, and (3) R&D-oriented
knowledge-base.

First, RTD is facilitated by *general conditions and resources*, including the
endogenous resource base within a region, its natural environment determining the
quality of life within a region, its industry base and market structures. Different
'hard' conditions trigger RTD as for example the existence of lead users in an established industry or the settlement of major multinational enterprises. 'Soft' factors
such as regional image and identity and the 'openness' of a region are both important inputs as well as outcomes of regional developments. One important process
in triggering regional RTD is that the regions and their key actors need to have
excellent regional 'antennae' in picking up and recognising external triggers, which
they can use to implement a regional RTD strategy.

Second, in terms of *institutional infrastructure* we distinguish between systemic,
individual, and process aspects vertically, and between macro, meso and micro levels horizontally. Regarding the institutional infrastructure, 'hard' systemic factors
refer to the overall network infrastructure required for RTD within a region, including political institutions on macro level, business intermediaries on meso level and
the general business support infrastructure on micro (firm) level. Individually, the
institutional infrastructure needs to be complemented by 'hard' factors such as key

Table 5.1 Stylised matrix of good practice elements in RTD

1. Input factors for RTD

General conditions / resources

Category	Hard factors	Soft factors
Territorial resources	Infrastructure, human capital, institutional capital	Regional image and identity
Market resources	Size, customer base, distribution channels	Openness of customer base for new processes, products, services
Industry resources	Age, size of industry base, technology orientation and level of technology use	'Curiosity', i.e. open for new ideas and divergence from routines
Processes	Regional 'antennae' in picking up regional triggers and using them in implementing regional RTD strategy	

Institutional infrastructure

Category level	Hard factors			Soft factors	
	Macro level	Meso level	Micro level	Macro level	Micro level
Systemic / organisational	Division of tasks and responsibilities between municipalities and other agencies	Dense institutional networks of intermediaries (chambers, business associations, unions, business support agencies)	Dense business networks Good general support infrastructure for entrepreneurship	'Open region'	High level of cooperation and interaction between actors
Individual	High communicators	Network promoters	Star scientists	Open minds	Networking skills
Process	Good governance • political commitment and coherence of institutional infrastructure • integration and openness at individual and institutional level Creation of social capital in the form of trust-based and reciprocal relationships within region				

Table 5.1 (continued)

R&D oriented knowledge base

Category level	Hard factors			Soft Factors	
	Macro level	Meso level	Micro level	Macro level	Micro level
Systemic	Existence of (semi-)public research infrastructure, universities	Existence of education and vocational training institutions	Special R&D support and education, instruments for research transfer	Existence of technical culture	Common values such as trust and reciprocity
Individual	Policies for attracting high skilled labour	Policies for upgrading skills	R&D policies, policies for upgrading skills	Attitude towards (new) technologies	Professional skills & social competencies
Process	Shift from individual and spatially dispersed learning to collective learning				
	Creation of technical culture				
	Creation of social capital in the form of trust-based and reciprocal relationships within region				

2. Critical factors

Over-embeddedness
- lock-in effects
- negative path-dependencies
- inertia

Possible indicators
- culture: traditional regional identity, often 'glorifying' industrial past
- technological regime and sector structure: lack of or low R&D-orientation, low technical culture and interest
- networks and firms within region: closed networks, focusing on intra-regional linkages, neglecting extra-regional and international linkages

Source: Welter and Kolb (2006).

actors like high communicators on macro level (political level), network promoters on the meso and star scientists on micro level. This is also reflected in the soft factors needed for the institutional RTD infrastructure to evolve, such as the openness of a region, the open minds and curiosity of actors, all of which are reflected in a high level of cooperation between different actors and good networking skills. Processes needed for improving or building such RTD institutional infrastructure refer to good governance within and amongst networks on different levels as well as the creation of trust-based relationships amongst different institutions and actors.

Third, RTD is influenced by the *R&D oriented knowledge-base*. People also matter with regard to regional knowledge, knowledge transfer and regional learning. In order for a regional R&D oriented knowledge base to emerge, a region requires a knowledge infrastructure on a systemic level, including research institutions and universities on macro level, educational and vocational training institutions on meso level and specific R&D support and education programmes as well as measures fostering research transfer on the micro level. Individually, knowledge might be attracted to a region by policies aimed at attracting highly skilled labour. 'Soft' knowledge-based factors include the existence of a technical culture on a systemic level and people's attitudes towards this as well as their professional and social skills and the existence of values supporting such a culture. All this helps foster learning processes within the region.

Malmberg (2003, p. 151) stated that there has been 'too much focus on interaction between firms within geographically defined spaces and numerous rather pointless attempts of trying to assess the degree to which there is actual interaction going on locally'. However, our analysis confirms that it is the interaction and *interplay of various factors with region-specific resource endowments* that will truly foster regional development. Therefore the next sections of this chapter explore the process of regional RTD, and introduce a model of how to foster effective regional interaction processes.

5.3 Methodology

This section describes the two critical phases used to develop the CRIPREDE Adaptive[3] Model[4] (i) transforming the 'matrix of good practice elements in RTD' into a working model, and (ii) developing, implementing and testing the model.

---

[3] The term *adaptive* in this context means that the model is not location, technology, industry sector, administrative, or infrastructure specific. The model has been designed in such a manner as to be applicable in any geographic or political domain as it takes the nuances of these domains into consideration through the process of vision generation, strategy development, action planning, and implementation.

[4] The term *model* in this context means 'process', a practical step-by-step methodology through which users can (a) audit and assess their current regional characteristics and attributes (both hard and soft), and (b) define and implement a vision, strategy and set of actions to transform their respective regions into more RTD oriented regions.

5.3.1 Transforming the 'Matrix of Good Practice Elements in RTD' into a Working Model

The underpinning philosophy of our research was phenomenological, using aspects of Grounded Theory (GT) (Glaser, 2002) as guiding principles in the process of developing the model. The Adaptive Model that evolved from this applied research was based on the interplay between the interviewing process of different cohorts of informants, the broader process of data collection and data analysis (Goulding, 2002). The process employed the sequence of interviewing,[5] developing theory, and re-interviewing in order to enhance the theory (Glaser, 1978; Glaser & Strauss, 1967; Lock, 2001; Strauss & Corbin, 1997).

The selection of the key informants (interviewees) was informed by Etzkowitz and Leydesdorff's (1997) *triple helix model* and Schramm's (2006) *entrepreneurial ecosystem model*. Therefore the interviewees were representative of enterprises (start-ups and large/multinational), higher level education institutions and government (local, regional and national).

Including entrepreneurs, micro and small to medium enterprises was an important aspect of our research as so often, when developing regional strategies, large, established firms are asked to contribute to the process. This is especially the case in emerging economies that are heavily reliant on Foreign Direct Investment (FDI). In such economies multinational enterprises (MNEs) are often included in the developmental process, and there is a tendency not to include indigenous SMEs or micro enterprises (O'Gorman, 2007).

Another important dimension to the development of the CRIPREDE Adaptive Model was to understand the role of developmental coalitions in the regional development process. It is the bringing together of a coalition of regional stakeholders (Ansell, 2000) and working on the 'soft field of interactions' among these interested actors and stakeholders where they can develop a network of linkages, dependencies, exchanges and loyalties (Sztompka, 1994), that leads to the conditions necessary to develop a 'habitat for the growth of a region' (Chong-Moon et al., 2000).

5.3.2 Developing, Implementing the Adaptive Model Process

The Adaptive Model was developed involving the six regions involved in the CRIPREDE project – City Triangle in The Netherlands; Cumbria in UK; Latgale in Latvia; Novo Mesto in Slovenia; Siegen-Wittgenstein in Germany; and South East Ireland. The industry mix, technological capacity, political structure, enterprise policy and support structure, general infrastructure, and the number of active

---

[5]In line with the flexibility and adaptability of the CRIPREDE Adaptive Model the interview process, techniques employed, and quantity of interviewees (regional stakeholders) varied from country to country involved in the CRIPRDE consortium. For further details see: Bugge et al. (2008).

stakeholders (regional players) are different in each of these regions; thus affording us the opportunity of developing and testing the flexibility and adaptability of the CRIPREDE Adaptive Model.

The development process had two linked aims: the parallel development of the Adaptive Model and Regional Actions Plans. This implied using a gradual process of scoping, focusing and continuous improvement involving regional stakeholders at each step of the process.

5.3.2.1 Developing the Adaptive Model

In defining the scope and focus for the Adaptive Model, the following were taken into consideration:

1. The model should be adaptive to enable use in regions with different characteristics;
2. It should enable identifying, influencing and measuring progress. In particular it should focus on options to address bottlenecks that prevent progress;
3. Valuable approaches for improvement should be extracted from (scientific) literature and experiences of well-performing regions;
   The model should be user-friendly and practical.

Based on theory researching previous experiences and practitioners' (i.e. regional stakeholders involved within the project) needs, the core of the model was focused on the interaction process in complex regional multi-actor settings ('soft' content). The model also incorporated elements of the process that needed to be linked to 'hard' (content) regional characteristics as well as objectives and instruments to influence performance. Furthermore, the regional stakeholders had clear opinions about how the model should be designed to influence regional innovation. Again the process perspective dominated and a question-based, decision-support process model was preferred. The design should facilitate discussion and simultaneously offer guidelines for development processes, decision-making and implementation. In particular, the model was designed to be applicable for use in multi-stakeholder settings. It was from a combination of this thinking, research and a range of guidelines for developing such a model (Welter & Kolb, 2006), the first complete draft version of the model was developed.

This first usable draft of the Adaptive Model was rigorously tested in Germany, Ireland, The Netherlands, and United Kingdom. Even across these four developed economy regions, there were many variants as to how the model was used to gather relevant data for each region. The feedback from regional stakeholders acknowledged the model's functionality. However, the regional stakeholders were less satisfied concerning the size (number, and degree of detail, of questions) of the regional audit part of the model. The following iterative improvements, involving research teams and regional stakeholders, resulted in consensus on 14 main questions (see Section 5.4).

Based on this feedback the model was improved, and retested in Latvia. In the Latvia test it was decided to specifically compare the use and applicability of the model in an emerging economy and a developed economy. The two regions used for this comparative analysis were the Latgale region, Latvia (emerging economy) and the South East region, Ireland (developed economy). Once again, the flexibility and adaptability of the model was demonstrated, as representatives of both regions were able to use the model effectively, taking their regional nuances into consideration during the process.

Even though the Latvia test was extremely successful, the model was improved further and tested again in Novo Mesto, Slovenia. In Slovenia the comparative analysis of the applicability and usefulness of the model in emerging and developed economies was the focus of the test. When completed, the Adaptive Model was considered applicable, usable and effective by stakeholders of all involved regions (see Section 5.4.1). The Adaptive Model was eventually released at an International conference in Ireland in October, 2007.

5.3.2.2 Developing the Regional Actions Plans

The ultimate test of the Adaptive Model was its use in auditing, and subsequent development of action plans for, all six participating regions. The task of the research teams (the CRIPREDE project partners) was, in all regions, limited to facilitating the process, providing information, and documenting and structuring results. The regional stakeholders performed the audits and subsequently developed the action plans. In each of the six regions, a mix of relevant stakeholders from public, private and research sectors were engaged in the process of regional action plan development, reflecting the 'triple helix' structure outlined in Section 5.3.1. In order to reconcile the demands of a verifiable and valid regional development process with the local vagaries of stakeholder governance, it was agreed that regional action plans would be developed using a pre-prepared template. This template allowed regions to express the key issues emerging from their audit and action planning process, yet allowed information to be compared between regions.

In summary, the Adaptive Model and Regional Action Plans were co-developed through a collaborative process where the research teams facilitated regional stakeholders in translating tacit local knowledge into structured results.

5.4 Results: Adaptive Model, Action Plans, and Improved Collaborative Processes

This close collaboration in developing model and action plans was not only viewed as a key part of the process within the participating regions; it was also acknowledged by stakeholders as an important incentive for improvements towards a sustainable regional development approach.

5.4.1 Adaptive Model

This is reflected in Figure 5.1, which depicts the relationships between model and process. The Adaptive Model combines three main 'building blocks'. Although the model should be viewed as a coherent 'package', the *Audit Tool*, as the 'core' part of the model, has been developed for stand-alone use. The *Guidance for Process Facilitators* and the *Strategy Development Tool* have been developed as illustrative examples and menu support for process design and decision-making, but regional stakeholders applying the model may choose alternative approaches.

Part 1, *Guidance for Process Facilitators*, applies a question-based, process management approach to assist regions in their process leading to regional action plans. The process is divided into six distinctive phases starting with attention to awareness and initiative and ending with implementation and efforts to ensure continuous improvement. The intermediate phases focus on 'bridging' process and content. This implies attention to interactive (strategic) planning and efforts aimed at building commitment. The descriptions of individual phases include information on inputs, activities, main steps, success and failure factors, and outputs.

Part 2, the *Audit Tool*, is a set of questions, divided into three categories, covering (i) characteristics of the regional profile, (ii) regional innovation processes and (iii) policy and instruments. The topics addressed within each category are:

(i) Characteristics of Regional Profile

1. Socio-economic development
2. Economic geographical characteristics
3. Living conditions
4. Culture
5. Presence of higher education and research
6. Structure of industry

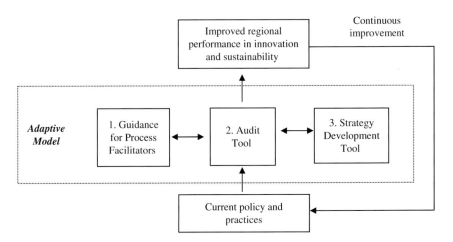

Fig. 5.1 Relationship between *adaptive model*, innovation, and sustainable development

(ii) Regional innovation processes

   7. Enterprises and innovation
   8. Access to information
   9. Role of public authorities
  10. Cooperation between stakeholders inside region
  11. Cooperation with stakeholders outside region
  12. Development of human capacity

(iii) Policy and instruments

  13. Regional innovation policy
  14. Presence and use of instruments

Each question asks for specific strengths and weaknesses. To facilitate this process, illustrative examples of potentially important aspects as well as a comprehensive list of sub-questions are provided. Finally, the user is asked to attempt to translate his/her overall impression into a score on a range of one to five, the highest number representing the maximum or most desirable score. The scores on the questions are plotted in 'spider web' diagrams, which serve as a basis for discussion.

Part 3, *Strategy Development Tool*, finally, offers decision-support for prioritising, a short feasibility check, and suggestions for actions that can 'fill the gaps'.

A valuable demonstration of the applicability of the Adaptive Model was that each region deliberately chose to perform its process in slightly different ways, thus applying the model in an adaptive manner. In particular the *Guidance for Process Facilitators* was used as a 'choice-menu': i.e. regional actors used elements based on their own knowledge of what would 'fit' into ongoing processes.

The *Audit Tool* was considered complete and coherent. It covered the essential topics of regional innovation. The audit questions, and in particular the 'spider webs', were viewed as helpful in facilitating a discussion leading towards identifying a strategy that would fit regional ambitions, needs and characteristics. The spider webs were experienced as useful for developing a relative scale concerning a region's weaknesses and strengths. On the other hand, regional actors expressed that they possessed insufficient information for scoring (bench-marking) relative to other European regions. However, this was not considered a major issue, because they generally were more interested in learning from good practices in well-performing regions than valuing any comparative ranking.

5.4.2 Regional Action Plans

The value of the Adaptive Model in facilitating collaborative strategy development was demonstrated in six different regional contexts involving a variety of stakeholders. The exact composition of these stakeholder groups varied from region to region, reflecting a number of differing factors.

Firstly, the stakeholder groups reflected the prevailing governance dynamic in that region at that time. In this respect, the group was time-limited and liable to change. In some regions (such as in South East Ireland), the process stimulated the development of a new forum for RTD development, which assumed a life of its own, existing beyond the project stimulus. In this case, it seems that the CRIPREDE process had provided the catalyst for some latent demand within the region. In other regions (such as Cumbria, UK, and Latgale, Latvia), a bespoke stakeholder group was brought together for the purposes of the project, although the regional action plan itself became incorporated within other regional development activities.

This latter point highlighted a second characteristic of the CRIPREDE action planning process; in many of the regions, the sustainability of the action plans, and therefore the sustainability of the actions themselves, was dependent upon the existing development architecture; the action plans became part of wider regional development strategies, and were presented to existing stakeholder groups and partnerships. In other words, the action plans themselves became embedded in the regional processes of economic growth and development.

Finally, one must take into account the differing context in six diverse European regions. National differences, such as the varying levels of regional decentralisation, the status and roles of universities in regional development, the standards and values associated with private sector engagement in decision-making, all had an impact on the composition and functioning of the stakeholder groups.

However, despite these differences the Adaptive Model enabled comparable interaction processes, leading to comparable, though unique in specificity and focus, outcomes. In some cases the resulting Action Plans were broad aspirational documents, designed at least in part to raise stakeholder awareness of the importance of RTD in a regional context. In other cases, they contained a menu of project ideas, which would collectively raise the RTD profile of the region. In general, seven principal types of actions could be discerned:

1. *Networking actions*, with the aim of increasing communication and interaction between key economic actors in the region. These kinds of actions were common to almost all action plans, reflecting the perceived importance of process factors in regional settings. For example, whilst the Stedendriehoek region (The Netherlands) set an objective to 'Increase regional decisiveness', South East Ireland set out to create a 'Spirit of Enterprise Forum' to draw together key actors

2. *Actions to promote the profile or brand of the region*, such as the target in South East Ireland of 'Creating a brand for the South East region'.

3. *Actions to enhance knowledge transfer and 'triple helix' partnerships*. These are expressed in different ways, such as Latgale's (Latvia) target of 'Facilitating academia – business knowledge transfer', Siegen's action of 'Targeting Further Education more towards Research and Technology Development', and that of the Stedendriehoek to 'Accelerate and improve knowledge transfer between applied universities and the regional entrepreneurs'.

4. *Actions to boost entrepreneurship*, such as Siegen's (Germany) objective of 'Fostering Technology Orientated Venture Creation and Cooperation for More

RTD on Enterprise Level', and Cumbria's (UK) action to 'increase the resourcing for, and availability of, dedicated innovation support to companies'.

5. *Actions targeted at specific sectors.* In many cases, these were sectors identified as having growth potential in the region, for example, the Stedendriehoek plan contained two actions to 'Develop a regional agenda for RTD on environment and energy', and to 'Use available opportunities for RTD in the manufacturing industry sector'. Cumbria's plan contained a more generic action to 'Support the growth of sector-based networks'.

6. *Actions to address shortcomings of infrastructure or the attractiveness of the region* as a place for investment. For example, Siegen-Wittgenstein set two actions for 'Enhancing the Attractiveness of the Region as a Living and Working Place' and for 'Offering Sustainable Industrial Sites for Future Demand'. Similarly, Cumbria set an objective to 'Enhance the attractiveness of the region as a place to do business'. The plan for Latgale in Latvia contained a generic action to 'Improve accessibility of the region', which was detailed through six priority infrastructure actions, perhaps reflecting the importance of infrastructure development in the new member states.

7. *Actions to enhance human capital or the skills base of the region,* such as addressing graduate retention (in Latgale and Cumbria), or compiling a skills database (SE Ireland).

The CRIPREDE Action Planning Process clearly provided a microcosm of bottom-up regional development. It demonstrated how local partners can be adequately facilitated in a collaboration process towards sustainable social and economic health of the region.

5.5 Discussion, Conclusions, and Implications for Policy and Practice

5.5.1 *Exploring Added-Value of the Adaptive Model Approach*

In the introduction to this chapter we referred to the key challenges of sustainability and the Triple-P principle in simple terms, ensuring that the needs of current as well as future generations are met concerning social, economic and environmental issues. Although nobody can know exactly what future needs will be, it seems quite realistic to assume that freedom of choice and general welfare will remain important. The challenge accordingly is evident. Significant differences in levels of welfare exist between regions across Europe. All regions therefore need to find their own unique, innovative way to improve welfare in a sustainable manner. However, as one of the regional stakeholders involved in the CRIPREDE project expressed: 'What has struck me most of all is the similarity of issues raised by many stakeholders in the regions constituting the CRIPREDE consortium' (CRIPREDE, 2007b, p. 36).

This indicates that although the issues, to a large extent, are similar, each region needs to find its own strategy depending on its identity, characteristics, resources, problems and ambitions. We will now discuss some aspects of the development process of Adaptive Model and Regional Action Plans under the following headings:

- Stimulates regions to assume ownership of improvement process
- Facilitates discourse in searching for best-fit between multi-actor setting and ambitions
- Initiates continuous interactive learning and development of adaptable modes of embeddedness
- Encourages pro-active governance

5.5.1.1 Stimulates Regions to Assume Ownership of Improvement Process

One of the most important observations was that the facilitating research teams always needed to adapt their approaches to fit ongoing processes. Each region had its own initial situation characterised by, for example, existing organisations, networks, plans and projects. Although this observation may appear trivial, its consequences for introduction of the model were considerable. A key to acceptance turned out to be emphasising that the Adaptive Model would enter, stimulate, and facilitate a process that was 'owned' by the region itself. This ensured that regional stakeholders would be able to maintain 'control' of the process and its outcomes: an issue that proved to be important because the process needed to open a discourse on roles and effectiveness of existing institutional settings as well as ongoing plans and activities.

5.5.1.2 Facilitates Discourse in Searching for Best-fit Between Multi-actor Setting and Ambitions

The auditing process, in particular, revealed that multi-actor collaboration concerning comprehensive regional policy on RTD was rather limited in all regions at the beginning of the project. Although a variety of working groups and networks were active, these usually covered only specific themes, sectors of industry, or involved only a limited number of actors. In other words there was a high density of existing institutional networks, but none that specifically addressed regional RTD in a multi-actor setting that included all relevant regional stakeholders. In that sense the functioning, and in particular openness to change, of regions and networks became an important issue. Generally, we experienced stakeholders in all regions as open minded and eager to explore new opportunities for improving regional performance. However, there seemed to be a widespread implicit assumption that the process of change preferably should be incremental.

The Adaptive Model approach intentionally leaves decisions to the regional stakeholders. On the other hand, in all regions it specifically stimulated and facilitated an open discourse on which institutional settings would fit the ambitions of improving regional RTD. In most cases establishing a common focus resulted in attempts to align, and adjust, activities of existing networks. Structural changes, on the other hand, were more diverse. In the 'Stedendriehoek' (City Triangle) region in The Netherlands initially a diversity of initiatives was limited to 'twin' relations of, for example, industry and authorities or industry and university. The Adaptive Model approach facilitated regional stakeholders in discovering that research, education, industry and authorities should collaborate. This resulted in an enlargement of the Regional Board of Economic Affairs. In at least two other regions, Siegen-Wittgenstein, Germany, and South East Ireland, the CRIPREDE-process even led to the development of completely new structural collaboration entities. It would be interesting to revisit these collaborations in a few years' time to see how sustainable those structures are.

Another characteristic of the process was that the number of key individuals involved in each region, often referred to as 'high communicators' or 'innovation promoters' (see Section 5.2), was quite limited. These individuals often already knew each other, and participated as representatives of industry or authorities in several networks on a diversity of themes. We argue that 'adopting' regional RTD and the Adaptive Model approach, regardless of how the topic was institutionalised, may have represented quite an attractive option to these actors. The project, as one of the regional stakeholders expressed, '[...] came to the region just at the right moment' (CRIPREDE, 2007b, p. 37). The importance of regional RTD for regional welfare is increasingly acknowledged, and the Adaptive Model provided an answer to their struggle, and often lack of know-how, in developing coherent strategies for this topic.

5.5.1.3 Initiates Continuous Interactive Learning and Developing Adaptable Modes of Embeddedness

The Adaptive Model therefore also matched 'the eagerness to learn and improve', so clearly expressed by several stakeholders. The model (see Section 5.4) is designed for, and intended to, stimulate and facilitate continuous regional learning and improvement processes as one key input to regional RTD. The process approach made each region use open questions as a basis for interactive discussions, and stimulated involvement of representatives of all relevant regional stakeholders. Besides leading to sometimes surprising audit results, this approach contributed to an improved mutual understanding concerning the different 'worlds' of the regional stakeholders. This included a better understanding of interests and objectives and the institutional context that governs action.

However, there was also some concern about possible 'protectionist' behaviour, which would imply that key regional stakeholders might have considered 'adopting' regional RTD into existing closed networks as a way to maintain control. By 'adopting' the issue within 'the establishment', some stakeholders would have considerable influence on decisions concerning strategies, allocation of means and actor involvement. Although no clear evidence of 'protectionist' behaviour was found, the fact that a relatively small number of key individuals fulfil important roles within regions, suggests caution about the existence, or development, of closed networks, and possible lock-in (see Section 5.2) and 'group-thinking', which could seriously stifle regional RTD learning, and development, processes.

Another, related, important learning effect was the acknowledgement of interdependency, which, again, underpinned the choice of applying interactive collaboration processes as the core of the Adaptive Model approach. Specifically this learning process changed the views on what innovation (RTD) is. At the beginning of the project, innovation in most regions was seen as something that happens within, and 'belongs' to, industry. Local and regional authorities generally fulfilled a facilitating, sometimes rather distant, role. The Adaptive Model, on the other hand, stimulates common development of strategies by stakeholders based on (potential) influence on regional innovation and general regional welfare and sustainability. A specific learning issue was therefore that regional stakeholders recognised that regional RTD could be significantly improved if authorities, industry and universities combine strengths and align strategies based on a clear picture of needs and opportunities to influence progress (i.e. the Triple Helix, see Etzkowitz (2008)).

A third learning issue concerned the interaction between the six regions involved in the project. The stakeholders from different regions certainly appreciated the possibility to engage in discussions and were all interested in learning from each other. The interaction not only led to the valuable recognition of common issues to be addressed, but also underpinned that it was necessary to be critical about attempting to 'copy' success stories from other regions. Success would always depend on adjustment to one's own regional characteristics and ambitions.

The issue of context-dependency was also encountered and quite pronounced in relation to the application of the model for auditing. Although the model worked adequately for individual regions, the regional stakeholders expressed that they had insufficient information for making any comparative (bench-marking) inter-regional assessment. This was clearly demonstrated during a workshop where two teams of stakeholders from different regions (and countries) arrived at final score-patterns that certainly did not reflect their relative socio-economic situations. A learning effect for the regional stakeholders was that there is a need for what we earlier (Section 5.2) have referred to as 'regional antennas' (also known as 'mavens'): individuals or organisations collecting and disseminating 'rich' information for their own region, based on a 'helicopter-view' and excellent extra-regional (preferably international as well) networks.

5.5.1.4 Encourages Pro-active Governance

The Adaptive Model is developed for multi-actor use. However, there still is a need to identify one actor that is willing to initiate and drive the process. Taking into account the 'natural' roles of different stakeholders, it was widely acknowledged that the regional authorities, in the future, should take on this role. The Adaptive Model process enables authorities to develop Regional Action Plans applying so-called 'good governance', which implies that other stakeholders are accepted, and integrated, into the planning process on a basis of equality and mutual respect. This approach creates a better mutual understanding of differences in ideas, interests and constraints given by institutional contexts, which, again, serves as an important stepping-stone towards increased trust: a key to *effective collaboration*.

Within the project, the national research teams involved as partners in the CRIPREDE project were responsible for facilitating this interesting development. The regions acknowledged the importance and complexity of such an interactive process, and specifically identified the need to involve an experienced, objective, process facilitator.

5.5.2 Conclusions and Implications for Policy and Practice

We, and our regional stakeholders, have experienced that the Adaptive Model process works. It facilitates, and stimulates, collaborative interaction leading to regional learning, innovation, and a more sustainable future. It fulfils the needs of stakeholders that want adaptive question-based decision-support, and no 'blueprint' solution, in their complex search for the 'right' strategy for regions with different characteristics. The approach thereby acknowledges real-life constraints of 'bounded rationality' (Simon, 1957), stakeholders need a process that enables decision-making under conditions of uncertainty and limited resources.

Specifically, the Adaptive Model challenged regions to make all regional stakeholders part of an interactive strategy development process, and stressed the need to adapt and learn within a continuous improvement helix. It addressed a Triple-P range of issues and stimulated discourse on choices, which explicitly link regional characteristics to RTD-processes and policy. The resulting Regional Actions Plans, important as they are, still only represent the visible 'top of the iceberg' of regional change that was initiated through the Adaptive Model, such as learning processes and change to networks, and interaction, in all regions.

However, there remains a question mark concerning continuity and embeddedness. The differences in regional resources and dynamics are large. Both issues may present serious threats to structural follow-up. The Adaptive Model includes, and stresses the importance of, a process towards continuous improvement, but it will be up to regional stakeholders to drive this process.

This leads to the implications for future policy and practice. A major challenge, in our opinion, is to continue along the road towards achieving an embedded, and simultaneously adaptable, real collaboration between *all* regional stakeholders. The key to success is to find the 'right' involvement of relevant stakeholders depending on ambitions, and possibilities: in particular the 'triple helix' networks of industry, authorities, and universities. This involvement should accordingly be 'tailor-made' based on clearly identified responsibilities, added value, and synergy. The collaboration needs to be pragmatic and at the same time have a strategic long-term perspective and framework. In other words, partnership-based collaboration in regional development should fit the needs of current as well as future stakeholders. In that respect, regional development can demonstrate true sustainability, in that it can encompass social and economic, as well as environmental, longevity. The search for such improved practices represents an important, and highly interesting, challenge for the future – for practitioners, researchers, and policy makers: through *constructive collaboration.*

References

Amin, A., & Thrift, N. (1994). Living in the global. In A. Thrift (Ed.), *Globalization, institutions and regional development in Europe* (pp. 1–22). Oxford: Oxford University Press.

Anderson, A. R., & Jack S. L. (2002). The articulation of social capital in entrepreneurial networks: A glue or a lubricant? *Entrepreneurship & Regional Development, 14*, 193–210.

Ansell, C. (2000). The networked policy: Regional development in Western Europe. *Governance, 12*(3), 303–333.

Bugge, K. E. (2003). Dilemma's bij Duurzame Revitalisatie van Bedrijventerreinen. *Proceedings of '2nd Industrial Life'* (pp. 43–50). Apeldoorn: TNO.

Bugge, K. E., Blokland, H., Lier, G., Peck, F., Mulvey, G., Hill, I., et al. (2008). *Developing knowledge-based regional economies.* Cork: Oak Tree Press.

Chong-Moon, L., Miller, W. F., Gong Hancock, M., & Rowen, H. S. (Eds.). (2000). *The Silicon Valley Edge: A Habitat for Innovation and Entrepreneurship.* Stanford: Stanford University Press.

CORDIS FP6 2009, What is FP6. (n.d.). Retrieved January 2009, from http://cordis.europa.eu/fp6/fp6_glance.htm#.

CRIPREDE. (2007a). *FP6-project: Creating a RTD Investment Policy for Regions in Emerging and Developed Regions.*

CRIPREDE. (2007b). *Developing competitive knowledge-based regional economies in Europe.* Booklet developed by the CRIPREDE-consortium containing brief descriptions of project, model, involved regions and, stakeholder opinions about the project.

Elkington, J. (1997). *Cannibals with forks: The triple bottom line of 21st century business.* Oxford: Capstone.

Etzkowitz, H. (2008). *The triple helix: University-industry-government innovation in action.* New York: Routledge.

Etzkowitz, H., & Leydesdorff, L. (1997). *Universities in the global economy: A triple helix of university-industry-government relations.* London: Cassell Academic.

Fromhold-Eisebith, M. (1995). Das 'kreative Milieu' als Motor regionalwirtschaftlicher Entwicklung. Forschungstrends und Erfassungsmöglichkeiten. *Geographische Zeitschrift, 83*(1), 30–47.

Glaser, B. G. (1978). *Theoretical sensitivity: Advances in the methodology of grounded theory.* Mill Valley, Calif: Sociology Press.

Glaser, B. G. (2002). Conceptualization: On theory and theorizing using grounded theory. *International Journal of Qualitative Methods, 1*(2). Article 3. Retrieved November 2008, from http://www.ualberta.ca/~ijqm.

Glaser, B. G., & Strauss A. L. (1967). *The discovery of grounded theory – Strategies for qualitative research.* London: Weidenfeld and Nicolson.

Goulding, C. (2002). *Grounded theory – A practical guide for management, business and market research.* London: Sage Publications.

Koch, L., Kautonen, T., & Grünhagen, M. (2006). Development of cooperation in new venture support networks: The role of key actors. *Journal of Small Business and Enterprise Development, 13*(1), 62–72.

Lawson, C. (1997). *Territorial clustering and high-technology innovation: From industrial districts to innovative Milieux* (ESRC Centre for Business Research working paper 54). University of Cambridge.

Lechner, C., & Dowling, M. (2003). Firm networks: External relationships as sources for the growth and competitiveness of entrepreneurial firms. *Entrepreneurship & Regional Development, 15*, 1–26.

Liao, J., & Welsch, H. (2005). Roles of social capital in venture creation: Key dimensions and research implications. *Journal of Small Business Management, 43*(4), 345–362.

Lock, K. (2001). *Grounded theory in management research.* London: Sage Publications.

Malecki, E. (1997). Entrepreneurs, networks and economic development: A review of recent research. In J. Katz (Ed.), *Advances in entrepreneurship, firm emergence and growth* (Vol. 3, pp. 57–118). Amsterdam: Elsevier.

Malmberg, A. (2003). Beyond the cluster – Local milieu and global connections. In J. Peck & W. Yeung (Eds.), *Remaking the global economy: Economic-geographical perspectives* (pp. 145–162). London: Sage Publications.

Maskell, P., & Malmberg, A. (1999). Localised learning and industrial competitiveness. *Cambridge Journal of Economics, 23*, 167–185.

Morgan, K. (1997). The learning region: Institutions, innovation and regional renewal. *Regional Studies, 31*(5), 491–504.

O'Gorman, B. (2007). *MNEs and new enterprise creation: Do MNEs have a direct impact on the amount of new indigenous high-tech start-ups in Ireland?.* Dissertation, Middlesex University, London.

Schamp, E. W., & Lo, V. (2003). Knowledge, learning and regional development: An introduction. In V. Lo & E. W. Schamp (Eds.), *Knowledge, learning, and regional development* (pp. 1–12). Münster: LIT.

Schramm, C. J. (2006). *The entrepreneurial imperative.* New York: HarperCollins.

Simon, H. A. (1957). *Models of man: Social and rational: Mathematical essays on rational human behavior in a social setting.* New York: Wiley.

Strauss, A. L., & Corbin J. M. (1997). *Grounded theory in practice.* Thousand Oaks, CA: Sage Publications.

Sztompka, P. (1994). *The sociology of social change.* Cambridge: Blackwell.

United Nations – World Commission on Environment and Development (Brundtland-commission). (1987). *Our Common Future,* United Nations.

Welter, F., & Kolb, S. (2006). How to make regions RTD success stories? *Good practice models and regional RTD. Beiträge zur KMU-Forschung* (2. Vol.). Siegen: PRO KMU.

Chapter 6
FOCISS for an Effective Sustainable Innovation Strategy

Jan Venselaar

Abstract Sustainable development will be a major driving force for future developments in businesses. Most companies are fully aware of that, but find it difficult to translate this insight into concrete actions. We have observed that companies, in particular small and medium enterprises, find it difficult to determine how sustainability can affect their business. The FOCISS (Focussing Innovation Strategy for Sustainability) approach offers that assistance. We have developed FOCISS in collaboration with enterprises from diverse industry sectors. By this approach the 'agenda' and the conditions for sustainable business of a company, in its specific situation, can be established. Stepwise, key areas of relevance, major issues therein and finally the most promising innovations, in view of economics and sustainability, are assessed. The approach uses primarily the views and expertise of the people working in the company, which improves the exchange of views and information and strengthens the collaboration on such issues through all parts of the company. A clear focus on sustainability can also improve the basis for collaboration with outside stakeholders. It strengthens the commitment and ambition to integrate sustainability in business strategy.

Keywords Sustainable business strategy · Small and medium enterprises · Systems theory · Stakeholders

6.1 Introduction

6.1.1 Sustainability and Small Enterprises

Recent management research literature discussing the relationship between sustainability, innovation and profitability shows a growing consensus that incorporating

J. Venselaar (✉)
Research Group Sustainable Business Operation, Avans University of Applied Sciences, PO Box 1097, Tilburg, 5004, The Netherlands
e-mail: j.venselaar@avans.nl

J. Sarkis (eds.), *Facilitating Sustainable Innovation through Collaboration*,   97
DOI 10.1007/978-90-481-3159-4_6, © Springer Science+Business Media B.V. 2010

sustainability and corporate social responsibility in the company strategy will strengthen the company's basis and profitability, certainly in the longer term (Bhattacharyya, 2007). Nevertheless Bhattacharyya concludes: 'The notion that corporate social responsibility (CSR) should benefit the firm as well is no news now. But a framework which guides managers so that they can decide which CSR initiatives make strategic sense to the firm remains elusive.' He observes that this missing practical framework is an obvious 'research gap'. Theory is not translated into practice. He concludes in his study that an effective 'CSR-strategy-filter' should be developed that is able to define those CSR activities which 'contribute to value chain activities and improve context of competitiveness'.

Surveys, often informal, of the various industry sectors show that most companies accept the necessity of sustainable development for society and economy and therefore is important for them as well. Our observations have established, however, that the majority do not have a clear view on how sustainability will affect them. In practice most companies, in particular those which are smaller, tend to act as if sustainability is important only for the larger companies and politicians. They consider sustainability more a risk than an opportunity and hardly acknowledge it as a major driving force of a future innovation strategy. In our studies we observed several causes for this discrepancy. A major cause is that sustainability appears to be too complex with too many issues involved. Most issues seem to have no direct bearing on the actual business and daily operations.

Companies select innovations based on short term considerations by looking at the existing markets and profitability under present conditions. They assume that these conditions will prevail long enough for the investments to pay off. Small and medium enterprises (SMEs) often do not have a clear strategic framework and have an insufficiently coherent vision of the future. Many businesses try to avoid risks by 'adhering to their trusted ways' so that radical innovations, often needed for future sustainable profitability, are not considered. Moreover, exchange of information on the importance of CSR and sustainability often stays confined to a small number of people within the organisation. The exchange of information between departments can be very limited, certainly in larger companies. Efforts to introduce new ideas are therefore not effective.

Sustainable management is often introduced and discussed by using one of many existing checklists. These checklists contain long and complete lists of aspects and issues that have to be considered. They are very useful when evaluating the sustainability of a company as a whole, or when an official report on sustainable performance is required (GRI, 2008). For SMEs the checklists are too impractical and even intimidating. These companies indicate that to motivate people and to make the checklists meaningful and useful for them, focus and priorities are needed (Bhattacharyya, 2007).

In order to achieve the required effectiveness of sustainable management and accountability every organisation would need to customise its system. The focus should be on those sustainability issues and activities, which are related to the actual core business (Hubbard, 2009). This reasoning also supports our observation that a company must concentrate on a limited number of priorities. Moreover, when sustainability related issues form a real concern for the core business and therefore future viability, they are more likely to create a 'sense of urgency', which stimulates continuing commitment to the changes and novel approaches needed.

6.1.2 The System Character of Sustainability

Because sustainable development is unthinkable without radical changes in the complex socio-economic structures that form our economy, a company must understand its own position, role and interests in it (Geels, 2002). It is therefore crucial to understand the system character of society, sustainability and the effect of such radical changes (transitions). It is widely debated, whether such transitions will take place due to economic drivers and market influences, or whether laws and regulations will be the major driver, but in either case they will take place. Companies, therefore, will have to respond to these changing economic and societal conditions and adapt in the right direction in order to survive.

A thorough understanding of these developments, its consequences and necessary actions is also essential to create an effective strategic cooperation with stakeholders. We observed in several cases, that companies became involved in discussions and collaborations on issues suggested by others, for instance in socially responsible business initiatives, which soon lost momentum because they were too broad and therefore, lacked the 'sense of urgency' for most participants involved. Companies must also be aware of the system level at which transitions take place and at which level their response is the most effective.

Three levels can be distinguished: the production level itself, the product chain and society and its socio-economic structures as a whole, as shown in Fig. 6.1 (Venselaar & Weterings, 2005).

The production processes of the company at the production level must be as clean and eco-efficient as possible. Relations with the own employees and community take place at this level. At the production chain level the company is a link in the material chain, from basic resource till waste/new resource and must be made as lean and eco-efficient as possible. The chain must be closed by minimising losses and optimising reuse, which requires information exchange and effective collaboration of the companies involved in the production chains. On the societal and socio-economics system level the company faces the challenge to respond to the changing needs and requirements resulting from the 'sustainable transitions' which take place in the systems and chains. Besides risks it offers new business opportunities.

All too often companies address only the production level, when discussing 'sustainable business'. In such cases it is common practise to emphasise 'Planet' aspects (environment, resources, ecology) and employee related aspects (labour conditions). The required actions are then mainly the responsibility of the environmental management department alone. Issues at the two other levels are the responsibility of the research, marketing and strategy departments and require intensive communication

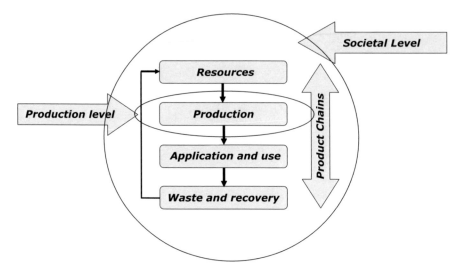

Fig. 6.1 Three levels of systems and transitions

and collaboration throughout the company. However, their lack of expertise on sustainability is a common occurrence. As a consequence the necessary measures regarding those levels are not understood nor taken.

6.1.3 Developing a Practical Tool

The complexity of sustainability is a major cause for the observed discrepancy between understanding the issues and actual behaviour of companies with respect to sustainability and for the lack of effectiveness of many actions taken. In the course of our studies and projects aimed at introducing sustainable business management and the necessary changes in strategy and practical operation, it became clear that new approaches are needed to assist and motivate (SMEs) in this matter. Existing approaches proved not to be sufficiently effective nor did they really motivate small companies to adopt sustainable business as an essential part of their business and innovation strategy. Bhattacharyya (2007) points out that a new approach must bridge the observed 'research gap' and be used effectively as the 'CSR-strategy-filter'. In response, we have developed a novel approach, which assists companies with the introduction of sustainable management and particularly with the recognition of their priorities for sustainable innovation. This approach is called FOCISS: Focussing Innovation for a Sustainable Strategy.

FOCISS is intended for SMEs (up to 250 employees). On the whole, they have neither the expertise nor the manpower to study sustainability and its consequences for them. SMEs furthermore indicate that studies should also not be too expensive or 'time consuming', at least not in the exploratory phase. To be efficient and effective it should be straightforward and fast and should lead to a practical business and innovation strategy, which allows immediate implementation.

Application of FOCISS furthermore allowed us to map the factors which influence the introduction of sustainable management in SMEs. For a number of companies we have evaluated which factors influence their choices and how these are accepted and implemented in their strategy, both in the short term and in the long term.

In order to study the effectiveness of FOCISS the views and innovation priorities of specific companies have been compared with their earlier strategic choices for innovation. For instance an important question is whether this approach has led to real and company specific issues and not just to generic, 'fashionable' and much publicised aspects, which are not necessarily the critical issues for that company. Some preliminary observations concerning these issues are discussed below.

6.2 FOCISS, the Principle Aspects

6.2.1 A Practice Based Approach

The FOCISS approach was developed through an empirical research method in which 10 companies were involved. The companies came from diverse sectors of industry: electronics, chemistry, food, construction of printing equipment, and housing. The starting point was a rough outline of the approach, which was further developed and amended on the basis of our observations and the comments of the companies. This process guaranteed that the most practical and effective approach would emerge. The various 'tools' needed are based on existing models and approaches, but were adapted on the basis of the experience and the comments of the companies.

The idea was to develop a general approach, which could easily be applied by different types of enterprises through minor adaptations in the model. The basic set-up of the approach was improved progressively, but not in a sector-specific way. Development is still continuing and will be based on our conclusions regarding its effectiveness in selecting truly critical innovations and stimulating actual implementation.

6.2.2 Basic Principles

In accordance with accepted principles a 'sustainable business' approach must address sustainability in its broadest sense: people, planet, prosperity (also called profit or 'added value') and, therefore, also the aspects and issues that are sometimes treated separately as 'corporate social responsibility' (CSR) and 'corporate governance'. They all contain relevant issues, which might prove essential for a company in terms of risks or opportunities. A good performance in one area of sustainability, for example in energy, does not guarantee future viability if critical issues are neglected in other areas. For instance we observed that some companies invest heavily in renewable energy, because at present that is perceived as the most urgent (and fashionable) priority when developing into a sustainable business. As a result, they present themselves as 'a highly sustainable businesses'. At the same time, however, those companies produce products and have activities which are not sufficiently sustainable. They are dependent on uncertain resources (copper, wood), on supply chains, which involve socio-economic problems with resources (bio-fuels, child labour) or generate products with adverse effects (obesity, difficult to recycle).

A sustainable business approach should not only cover the total production chain for a specific product/activity but also the larger 'systems' of which the company is a part, as discussed in Section 6.1.2. Issues, risks and changes in any stage will inevitably affect the company's products and activities. One should therefore zoom in on the actual role and interest of a company in specific socio-economic systems and the way sustainability driven system transitions might affect these (for example, scarcity, new product demands, and novel technology). Specific aspects and areas within those transitions might offer crucial opportunities and risks for a company's continuity. A company must recognise those critical issues and find the best approaches / innovations to stay profitable in the future (see Fig. 6.2). All too often companies feel that they cannot control those changes and tend to ignore them.

Because sustainable development will be the major and inevitable driver of the economy, a sustainable strategy has to be adopted before choosing innovations. However, as we observed earlier, common practice is to first select an 'interesting' innovation based on (usually short term) financial and market factors. Sustainability considerations are only brought in as a second step with the intention to make the selected innovation as sustainable as possible. However, such an innovation doesn't necessarily fit into a sustainable business strategy leading to a sustainable and viable future for the company.

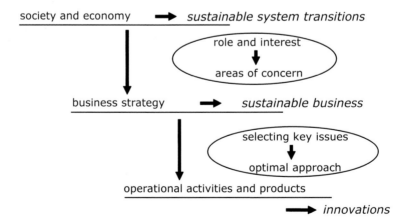

Fig. 6.2 Stepwise zooming in of the FOCISS approach

An innovation selected to address only one key sustainability issue, is not automatically sustainable either. Sustainable housing, for example, must also imply attention for transport and commuting to work or social problems. A renewable resource might create social and ecological problems during production, as is presently the case with the development of 'first generation' bio-based fuels using food grade vegetable oils and maize. Testing for such adverse effects is therefore a prerequisite for any attempt to implement innovations in the strategy of the company.

6.2.3 Creating Commitment and Collaboration

The results of a study must have follow-up actions in which the conclusions are incorporated in the business strategy, actual steps are taken to change business operation and introduce the selected innovations. Creating a positive and stimulating attitude to take follow-up actions are to part of the approach.

This process implies better appreciation of what sustainable development and sustainable business really entail, the actual role a company has and the weights assigned to sustainability aspects by the company. An inventory and assessment of opportunities and risks based on the larger sustainable developments such as climate change and growing world population and the transitions which might be the result of that linkage (hydrogen economy, more recycling) is a prerequisite.

The economic consequences should be clearly established. It must be emphasised that the selected options should not only be profitable at the present time, but should also form the basis for future strength. It must be clear that this is not necessarily a contradiction as commonly perceived.

The entire company staff must be involved in the study to create a sound basis for communication and collaboration on these issues. This involvement also creates a more effective commitment through the whole company. Such commitment and collaboration is further strengthened when the conclusions are based on their own experience and views. Therefore, the input information in FOCISS should primarily be based on the expertise, views and ambitions of the staff of the company. External input and additional research are needed to study specific items and to work out details of innovations. It is critical that the decisions are made by the staff and the management itself, mainly on the basis of their own views and expertise.

6.3 FOCISS, the Practical Aspects

6.3.1 The Set-Up

In principle FOCISS is a protocol which zooms in on key aspects and priorities in five consecutive steps. In the first step the scope of the project, in terms of products and activities, is defined. The second step zooms in on the position and role of the company with regard to sustainable transitions. The third step focuses on the key areas of sustainability which are relevant for the company. The fourth step identifies the major issues within these key areas. Finally, in step five the most promising innovations with respect to the various dimensions of sustainability are selected. The 'zooming in protocol' is structured in such a way that it leads to a significant reduction in time and effort a company and its advisors have to spend. Some tools are incorporated for structuring interviews, discussions, reflection and selection. These are specifically designed or are existing tools adapted for this purpose. This stepwise zooming in protocol is summarised in Table 6.1. In practice, strict adherence to this protocol has indeed proved to be necessary for obtaining reliable results at the required speed and efficiency.

A thorough introduction into FOCISS is necessary to inform and involve all people in the company, in particular those who are directly participating through interviews and meetings. It must clear the ground for an open discussion on what sustainable development really is and what it implies for a business. Misunderstandings, false assumptions and clearly biased views have to be avoided.

Conscientious involvement of all decision and opinion makers in the company is necessary. Because of time constraints only a limited number of these people can be interviewed but they can give their information and views at least in the discussion meetings. The involvement of other stakeholders is also important, especially external stakeholders that may collaborate with the company. Their information and views may contribute to various insights and add value to the decisions made. Moreover, early involvement facilitates their commitment to these decisions.

The scope of the sustainability study must be precisely defined. The subject of the study must be bounded and could be one product or a well-delineated set of products or activities. When combining products and product chains, they have to be sufficiently coherent and comparable. When not comparable, selection of key issues

Table 6.1 Outline of the FOCISS protocol

1	Preparations	Introduction in company
		Defining scope and selecting participants
		Collecting background information
2	Place of the company and generic issues	The systems on different levels and production chains involved and the major developments in those
		Specific issues in the industry sector, on that location
3	Key areas of attention	Interviews
		Discussion and selection of key areas
4	Key issues	Inventory and elaboration of issues named for the selected key areas of attention
		Discussion and selection of key issues
5	Sustainable innovations	Inventory of options for change and innovation
		Rough estimates of costs and (future) profits and 'sustainability effect'
		Discussion and selection of key innovations and starting points

and priorities is almost impossible. For instance office copiers which are leased involve other key issues than small copiers that are made for the consumer market.

An inventory of major trends in sustainable development, which take place in a particular industry sector, should be made leading to the so-called 'sustainability mirror'. Those have to be transformed into a recognisable picture of 'sustainable development trends' for the company involved. For instance the larger issue of climate change is translated in more practical issues as levies on CO_2 emission and emission trade, restrictions on energy use in general, options for renewable resources as well as new markets for substances and materials used for other forms of power generation and low weight materials for reduction of energy consumption. With the help of these practical issues they can recognise the actual effects for their businesses.

6.3.2 Interviews and the FOCISS Matrix

The interviews would be held with a limited number of persons. Selection of inter-viewees should keep in mind the various aspects and issues and expertise of the individual. It might be necessary to interview representatives of external parties. The people involved should be strongly encouraged to give their personal views and con-tribute on all issues of their expertise. These external parties may have more insight then those within the company. To create an atmosphere in which people feel free to express their views, the interviews should be a one-to-one basis. Key areas, issues and innovations are selected from the information obtained during those interviews.

An interview matrix is developed creating a clear overview of all aspects that have to be reviewed over the whole production chain and for 'People, Planet, Prosperity': called the FOCISS matrix (Fig. 6.3). The matrix is used to stimulate people to express their views and use their imagination. Working systematically through all the fields, the risk to overlook relevant aspects or issues is reduced.[1] It has been based on a matrix developed specifically for environmental issues, but also used in various other methods (Leopold, 1971). On the horizontal axis all the stages of the total production chain are listed. Since these can differ from company to com-pany, they have to be adapted for each individual case. Usually, for convenience the number of stages is restricted to between seven and nine stages. On the vertical axis the various sustainability aspects (People, Planet, Prosperity categories) are clus-tered into 12 groups. The total number of 'sustainability sectors' to be inventoried by means of the matrix is on average about 100.

Usually four to five key persons are interviewed in depth using the matrix. These give information and views on all possible aspects and issues that might be of inter-est for a specific field in the matrix. They are also asked to rate the issues mentioned. The number of interviewees might differ, dependent on the size of the company.

---

[1] E.g. the HAZOP (Hazard and Operability) analysis method for process safety issues operates in the same manner (Lawley, 1974).

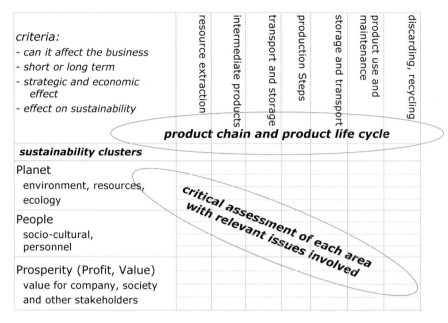

Fig. 6.3 Basic outline of the FOCISS matrix, used for the interviews

Interestingly, in our application of the methodology, we found that a larger number of interviewees did not result in many more issues or more precise ratings.

Issues involve risks and constraints with respect to continuity, sustainability and profitability, but also business opportunities and new options for better performance. The information is collated and discussed in a meeting with the staff and other parties involved. Based on this inventory, three or four key sectors (= matrix fields) are chosen.

In the following stage the issues from those key sectors are described and inventoried in more detail. In a second meeting subsequently three or four key issues are then selected. In the third stage the possible innovative approaches to handle those key issues (risk or opportunity) are inventoried and described. A first evaluation of their economic effect and their effect on the sustainability of the business is made. On the basis of that result, the most useful innovations are selected in a third and final discussion meeting.

6.3.3 Rating Method

To facilitate priority setting, the persons involved are asked to provide an importance rating for each of the aspects and issues in the various steps. Several rating methods have been tried, from simple to rather complicated methods. Forced rating proved to be the most practical. The results of the interviews are combined and reported in the form of median values and differences between the highest and the lowest rating given. The fields with high median values and those with a large scattering including high values should be targeted for discussion in the meeting. In practice there is usually a series of 10–20 fields clearly outlined. Fields with lower scores are not further discussed. This focussing needs to be approved by a consensus of the participants. Rating is rather subjective of course, based on personal information, views and expertise. Combining the results however creates a sort of 'balanced subjectivity'. These results do not automatically determine the priorities, but form the basis of the discussions in the meetings.

6.3.4 Structured Discussion and Selection

Each 'zooming in' step is concluded by meetings with strongly structured discussions. They must create an effective exchange of information and views and lead to a focus on the most relevant points of discussion. A two stage rating process is applied, again based on 'forced rating'. The first rating is given before any discussion has taken place followed by a second rating at the end of the discussion. Between the ratings the reasons for the individual ratings are discussed. In this manner information and views are exchanged. The last rating determines the final selection of respectively the key areas, the key issues and the most sensible innovations.

Those who have been interviewed participate in the meetings. Preferably other staff members who have views and additional information on the issues that have emerged from the interviews are also included in the more complete meetings. It also serves to strengthen their involvement and commitment. It might be useful to involve external stakeholders with which the company has to collaborate at later stages of the process.

6.4 An Evaluation of FOCISS Effectiveness

6.4.1 Benchmarking

A preliminary effectiveness study of the FOCISS approach has been completed. Further analysis is certainly needed. In assessing the approach, the following questions should be answered: 'are the selected key areas, key issues and innovations, indeed the essential ones' and 'are essential areas, issues and/or innovations overlooked'? A comparison with the results of other methods would be the optimal method. This step proved to be difficult. To our knowledge truly comparable approaches do not exist, certainly when it concerns the other aims such as creating commitment and amount of time and effort involved.

Benchmarking was completed, to a limited extent, by including companies that have been involved in earlier programs on sustainable management. Luckily, many companies who have gone through similar exercises exist within the Netherlands. FOCISS prioritised mostly the same key areas and issues, but also helped identify new emergent areas and issues. These results confirmed that FOCISS is able to generate a broader view than just the obvious. Moreover, these companies felt that their present priority setting was strengthened by this approach.

Another concern could be the availability of all relevant information within a company itself. To date, only in one study were customers and suppliers invited to participate. In that study they were crucial in the design and construction of the product which was subject of the FOCISS study. Companies appear to be rather insecure about discussing such issues with 'the outside world'. However, some comparisons could be made with information available in reports from (non-governmental organisations) NGO's and governmental bodies. They indicated, although more generic, the same issues for the particular industry sectors as in our studies. But 'unique issues' could not be checked in this way. However, representatives from environmental organisations (NGOs) and others confirmed the specific issues identified by the FOCISS approach. Therefore we feel rather strengthened in our view, that in general people in companies have a rather complete overview of the issues and problems, which exist in relation with their activities. It is just not information people easily volunteer nor discuss in 'normal' business meetings. FOCISS, in particular the individual interviews and discussion meetings, create an opening for that.

6.4.2 Better Selection

This first assessment of results and characteristic outcomes shows, that in nearly all company studies a limited number of key areas and issues could be easily identified. The rating method resulted in a very clear division between high and low ranking items. The selection of the typically three or four key areas, critical issues and subsequent optimal innovations, seldom proved difficult. Discussions tended to converge quickly on the obvious critical key areas and key issues.

Key critical issue results can be divided into three groups: 'expected', 'to some extent unexpected' and 'complete surprises'. The expected and obvious areas and issues are those which are very recognisable and fashionable and are often already being dealt with in a company. They immediately scored high ratings in the interviews. These items were primarily concerned with their own production processes, better efficiency, less energy and reduction of waste. Nevertheless, such issues did not necessarily end up in the final list of priorities. In the discussions their importance to the company proved to be overrated in many cases.

The 'to some extent unexpected' areas and issues typically concerned socio-economic developments in the region or 'elsewhere'. They are generally known in the company, but were not initially seen as crucial, but received a high score in the discussions. An example is the availability of materials and intermediates which are imported from politically unstable areas or which are manufactured by 'unethical production practices'. In earlier studies they were considered as less relevant, since the company felt that it was not in a position to improve the situation.

Totally 'unexpected' areas and issues, that no one else was really aware of, were often brought up by just one individual or suggested by the analysts. Those are the issues that are often 'unique' and originate from the specific processes and activities, or the particular circumstances in which a company has to operate. One example is the effect of animal diseases for a 'chemical company' which uses animal material as a basic resource. But also packaging is such a case, which most people in the company had seen as someone else's problem.

Understandably the issues and innovations for the 'non obvious key areas' caused the most debate. Proposals to choose priorities in those areas were met with reluctance. Not because one was uncertain about the priority it should receive, but because the obstacles to address them were clearly seen. It will cause a fundamental change of course. In particular research and investment programs would have to change focus. Any attempt to implement these in an organisation with existing but as yet non-sustainable priorities, are expected to cause frictions, in the organisation, but also with suppliers or customers.

6.4.3 More Fundamental Innovations

In only half of the studies the final step of selecting critical key innovations or starting points for making this selection was reached. Some companies were already quite satisfied with key issues on which further strategy development could take place. In some cases new investments were already planned, so these issues would be pursued 'automatically'. Sometimes the complexity of the key issues did not allow for 'easy' decisions to be made and more study proved to be necessary. Sometimes the choices for innovation were totally obvious and not very complicated to implement.

In most cases innovations were selected that involved simple alterations addressing specific environmental and social issues. These innovations did not have a large impact on other issues or sectors and were confined to the lowest system level i.e. production within a company (see Section 6.1.3). It could be, for instance, a more efficient separation, a new process based on a different recipe, specific environmental measures and a new supplier. These measures were easily identified, the solution was readily available and relatively easy to implement.

Innovations that require changes in the total product chain (system level two) were selected less often. They concern changes in the way the chain operates, and require 'integrated innovation'. One example concerned a new raw material which led to new processes and somewhat different products. Another example was the attention for socio-economic factors when buying low-priced materials, which are produced under disreputable conditions and could lead to adverse reactions with NGO's and in due course with customers. This category often contains ideas that already exist within the company but are difficult to implement and therefore low on the priority list. Usually the obstacles and risks are considered to be too large, certainly when totally new technology is involved. In most cases innovations from this category are selected when external pressure leaves no choice. These are the innovations for which collaboration with external partners, in particular suppliers and customers, is essential, which adds obviously to the complexity in implementing them.

A last category of innovations concerns the 'revolutionary changes' in the way the company operates, in its products and/or in the way it helps society 'to take care of its needs'. Usually drastic changes are not immediately required, but any future change and innovation will now have to fit in the direction dictated by such a 'third level innovation'. In some studies such innovations were discussed, but finally not selected, because they were considered too difficult to implement, at least for the present.

6.4.4 Reducing Obstacles for Implementation

Organisational aspects often prove to be obstacles when selecting the most suitable innovations for future viability. When selecting those innovations compromises are made, often implicitly. Evaluation of the selection procedure showed this to be the case especially for critical issues and innovations that were 'unexpected'.

An evaluation of the arguments used, showed that assessment of the impact and the complexity of innovations is determined by two main factors. One factor is the system level involved in implementing an innovation, meaning that changes have to occur there and consequently collaboration on that level is needed. At a higher system level more changes are needed and the influence of a single company is smaller. The other factor is the complexity of the changes in the company, its processes and organisations, that are needed to introduce an innovation. Adding equipment and procedures or exchanging one for another that is more efficient, is simple. Introducing a complete new process which requires new equipment, new training of operators and new organisational structures in a company is very complex. It entails high risks, certainly for SMEs.

To assess that problem, a simple diagram with these two 'complexities' along the axis is introduced (see Fig. 6.4). It shows major areas of issues and innovations for the companies involved in the projects which pertain to the chemical industry. Most of the innovations preferred by the companies tend to be positioned in the lower left part of the diagram. It is also observed, that the better options for sustainability and viability are found higher on the right side. Especially smaller companies, however, see (probably justly so) too many obstacles to enter that region. It forms an

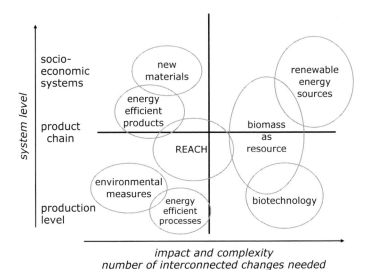

Fig. 6.4   Impact and complexity of issues and innovations in chemistry

essential factor in the often mentioned 'innovation paradox', particularly for SMEs (Venselaar & Weterings, 2005).

6.4.5 Improving Communication, Commitment and Collaboration

During the evaluations, which were held after completion of the studies, nearly all companies confirmed that sustainability had never before been discussed on such a broad platform within the company, which was looked upon as the main advantage of FOCISS. They also agreed that solutions were found that could not have arisen without this mode of communication. They also perceived that views and conclusions generated in this way are felt to be their 'own'.

As yet there has been no evaluation what the effect of the studies has been after a longer period. We have observed, however, that two companies, after completing their FOCISS study, have become leading partners in a regional program which has the aim to introduce sustainable business in their industry sector. Individual contacts with others have confirmed that companies are 'still working on it' but the precise status of the efforts is not known. A formal evaluation is planned.

In principle, companies agree, that collaboration with outside partners is easier when it is clear which concrete issues and priorities a company wants to address and discuss. Outside partners are for instance NGOs and authorities or suppliers and customers that are starting their own sustainable business strategy. They address them already on a broad range of specific issues. However a company often does not know how to react to that and how to determine risks and opportunities when responding to such address. Stakeholders such as NGOs and regional authorities, with which we cooperated in projects, indicated indeed that such a 'focussing' approach would be effective in stimulating collaboration. Their opinion is that limiting the number of concrete issues concerning sustainable development results in a more practical collaboration with companies. It is easier to involve companies in projects with the same priorities, instead of a broad introduction of sustainable management in all fields.

Furthermore, collaboration with suppliers of knowledge, such as universities and technology institutes, can be improved. Much knowledge is offered, but companies often do not have a clear view on which knowledge they require. When priorities for a company are clearer it is easier to define needs and requirements for cooperation.

6.5 Conclusions and Recommendations

6.5.1 General

A limited number of priorities and factors for a sustainable business and innovation strategy are shown to be most effective. FOCISS is a useful tool in this narrowing down of priorities. Application of FOCISS has occurred with a number of companies. The companies involved agree that the approach led to a selection of essential sustainability issues and innovation ideas over a broader range of areas than they had previously considered. Moreover, they acknowledged that it resulted in actions that they indeed recognise as vital to its core business. It was also obvious that developing a sustainable business strategy was less complex than appeared at first sight.

With regard to the development process for the FOCISS approach, direct collaboration with companies, which need this tool, is stimulating and also necessary. Further development of FOCISS, based on a more in-depth evaluation of the long-term effects for the companies that were involved, is certainly required.

6.5.2 Business Management

Communication within a company is shown to be of crucial importance when introducing a sustainable business strategy. Communication is essential for selecting the critical issues and the optimal innovations, but also for creating a sense of urgency and ownership of the results.

The exchange of information and, ultimately, collaboration is essential, not only between the disciplines and departments in a company, but also with other companies in the product chain. It is not probable that one individual has the complete information required to effectively evaluate all issues a company can encounter. When selecting new fields of innovation, evaluating constraints and setting priorities, all actors need to be consulted. It should be common practice to involve outside stakeholders in the process determining the key issues, which is a common recommendation in all literature on sustainable management. In practice most companies are very reluctant to do so. Up till now, usually outside stakeholders are consulted only on isolated activities and specific issues, which are not directly relevant for the future viability of the company. The impact of outside opinion and actions is underestimated.

Collaboration with external partners is underestimated in the same manner. Clear priorities, underpinned with arguments, which are fully implemented in the strategy of the company, facilitate cooperation and selection of external partners. With a limited number of issues and options and a sufficient insight in their importance, discussions can be fully productive. Collaboration with external partners also facilitates implementation of 'non obvious innovations' which, from a strictly internal point of view, appear less attractive and too complex.

6.5.3 Policy

Sustainable development and in particular transitions in chains and systems are facilitated by intensifying collaboration between the companies in a product chain. We have shown that effective collaboration is stimulated, when the specific issues, opportunities and roles of the companies involved, are specific and when these companies recognise their own priorities.

National and regional authorities could strengthen their present policies regarding sustainable development, by stimulating such priority setting and subsequent collaboration between companies on those issues. Specifically, it should be part of their programs and demonstration projects, where they challenge companies to adopt better practices for sustainable business management. As shown in Fig. 6.4, innovations that are complex will hardly be supported by individual industries on their own, in particular when it involves SMEs.

6.5.4 Research

The above conclusions provide a clear message for research and also knowledge institutes. The question is whether research activities are sufficiently directed to the needs of the companies and society. All too often newly developed knowledge and technology are 'sold' as 'being sustainable on its own'. It is shown that focus should be on innovations that contribute to the long-term sustainability of companies, not on innovations that are sustainable as such. Furthermore, a better focus of research on these aspects would solve the innovation paradox to some extent (Venselaar & Weterings, 2005). Institutes should not just offer knowledge, but should attempt to become involved with a company and groups of companies in a production chain at an early stage, during which the roles of the different partner and their priorities for sustainable business are identified. Only by such early collaboration, the actions and knowledge that fit the requirements and needs of the industry in that situation, can be defined and made available. Only then knowledge will be translated into effective innovations leading to real sustainability for companies and finally society.

Acknowledgments The initial development of the method (under the name DOSIT) received a grant from the Province of Gelderland and was carried out in 2004 and 2005 by a consortium, which consisted of TNO (the Netherlands Organisation for Applied Scientific Research), the University for Professional Education Arnhem-Nijmegen (HAN), Avans University for Professional Education (Tilburg) and Tertso Innovative Pathways for Environment and Sustainability[2] (Berendsen, Ansems, Appelman, & Venselaar, 2006). The subsequent development into FOCISS was carried out by the Avans Research Group Sustainable Business Operation as part of its research program with the purpose of assisting SMEs to introduce effective and profitable sustainability in their business operations. The authors gratefully acknowledge the contributions of the various partners, companies and the students and in particular John Hageman who coordinated part of the research and prof. Tom Reith for his comments on this paper.

References

Bhattacharyya, S. S. (2007). *Development of a CSR-strategy-framework.* Congress papers Corporate Responsibility Research Conference 2007, University of Leeds, UK, 15–17 July, 2007. http://www.crrconference.org/downloads/crrc2007bhattacharyya.pdf.

Berendsen, G. J., Ansems A. M. M., Appelman, W. A. J., & Venselaar, J. (2006). Profit for sustainable business, the DOSIT approach (in Dutch). *'Kwaliteit in Praktijk'*, June 2006, Kluwer Deventer (NL).

Geels, F. (2002). *Understanding the dynamics of technological transitions. A co-evolutionary and socio-technical analysis.* Enschede: Twente University Press.

GRI. (2008). *Global reporting Initiative; the 2008 update of the reporting guideline.* http://www.globalreporting.org/ReportingFramework/G3Online/.

Hubbard, C. (2009). Measuring organisational performance: Beyond the triple bottom line. *Business Strategy and the Environment, 19*, 177–191.

Lawley, H. G. (1974). Operability studies and hazard analysis. *Chemical Engineering Progress, 70*, 45.

Leopold, L. B. (1971). A procedure for evaluating environmental impact. *Geological Survey Circular No. 645*, US Geological Survey, Washington, DC.

Venselaar, J., & Weterings, R. A. P. M. (2005). Sustainable development in chemistry, engineering and industry, spontaneous transition or innovation paradox?. *Congress papers, World Congress Chemical Engineering (WCCE) 2005, Glasgow.*[tertso]

---

[2]the author worked at that time for all four institutions partly as adviser, partly as professor Sustainable Business Operation
[tertso]indicates that the publication is available as PDF on www.tertso.eu: look under 'publicaties'.

Chapter 7
The Emergence of Sustainable Innovations: Key Factors and Regional Support Structures

Peter S. Hofman and Theo de Bruijn

Abstract This chapter analyses the emergence of sustainable innovations in a selected number of firms and addresses key explanatory factors that contribute to emergence and diffusion of the innovations. The focus is particularly on regional support structures that facilitated the innovation processes, and on gaps between the needs identified within firms' innovation processes and functions provided by support structures. Ten sustainable innovation processes are analysed to gain insight in the relationship between the nature of the innovation process, the type of needs for firms, and the type of functions provided in regional innovation systems. It is concluded that especially for small and medium-sized enterprises (SMEs) demand articulation remains a major barrier as users are often only involved when the innovation is ready to enter the market, while regional support functions in this respect are deficient. Moreover, SMEs have major difficulty interpreting and antic- ipating sustainability policies and regulations at local and national levels, leading to innovations that face major regulatory barriers or are unable to cope with policy changes.

Keywords Small and medium-sized enterprises · Regional innovation systems · Radical innovation · Innovation support structures · Policy change

7.1 Introduction

Innovations are often viewed as technological phenomena. While technology plays an important role in many innovations in most cases collaboration and manag- ing information resources are as important. Many firms lack the capability to innovate on their own. Rather, it is the alignment of knowledge, perspectives and financial means from various actors that determine the successfulness of an

P.S. Hofman (✉)
Nottingham University Business School China, Ningbo, 315100, China
e-mail: Peter.Hofman@nottingham.edu.cn

innovation. Collaboration is, therefore, a crucial element in creating innovations. In innovation processes, entrepreneurs, start-ups and established firms interact with various organizations in the process of generating ideas, building innovative proto-types and bringing innovations to the market. The capability of a firm to innovate is therefore not solely determined by internal factors, such as its strategy, culture and organization, but also by the nature a firm's interaction with external actors, such as knowledge institutes, government organizations, users and capital providers. These interactions are shaped to some extent by firms themselves, but also significantly by the nature of the innovation system in which they operate. Innovation systems give structure to the interactions of various organizations involved in innovative processes. The national systems of innovation approach emerged as an explanation for the fast rise of Japan as an industrial power (Freeman, 1987). Crucial for the success of Japan's economic growth has been the ability to organise, mobilise, and direct efforts of a range of actors such as industries, research institutes, educational organisations and financial institutes along strategic visions set out by government in interaction with research institutes and industries (Freeman, 1988). In the inno-vation system discussion, Lundvall stressed the importance of interactive learning, for example between users and producers (1988), and focused on elements such as trust (and the formal institutions behind it) and mechanisms of exchange of tacit knowledge (based on skills, experience, and routines) in innovation processes (1992, Lundvall, Johnson, & Andersen, 2002). Next to national also regional innovation systems matter. Explanations of success of successful regions, industrial districts and regional clusters, focussed on the importance of interactive learning, key roles of processes of information dissemination and knowledge diffusion of local private and public organisations and roles of informal networks and social capital (Dimitriadis, Simpson, & Andronikidis, 2005; Morgan, 1997). In the case of sustainable inno-vations this is even more important as sustainable development requires radical innovations next to more incremental changes (De Bruijn & Tukker, 2002; Hartman, Hofman, & Stafford, 1999, 2002). These radical innovations assume reorienting organizations, processes and products. This goes beyond the capacity of individual firms. For sustainable innovation systems the importance of building some kind of collective vision is stressed, such as regions that develop an integrated sustainability vision and focus on developing sustainable solutions in particular areas of strength (Gerstlberger, 2004). A topic that will be adressed here is what regional support structures are needed to support sustainable innovations and to what extent these differ from regular regional innovation support. The approach in the paper will be to analyse a number of sustainable innovation processes and identify key needs for support.

The analysis focuses on a number of key questions:

– What are the suppport needs of firms in sustainable innovation processes?
– How do these needs differ across types of firms and innovations, e.g. small and medium-sized firms (SMEs) vs large firms and more incremental vs more radical innovations?
– How effective are regional support actors in providing functions in sustainable innovation processes?

The paper starts with a discussion of the nature of sustainable innovation and the type of actions required to bring these innovations to the market. This is followed by an overview of key functions that may be provided to innovating firms by support actors within regional innovation systems. A following section analyses a set of sustainable innovations to understand the need of firms for support and the way regional innovation systems were able to provide those needs. A final section draws some conclusions.

7.2 Sustainable Innovations and Firms' Needs

Innovations differ in how radical they are. The most common form of innovation in firms is incremental, and builds upon existing competences, technologies, functionalities and market linkages (Abernathy & Clark, 1985). For sustainable innovation incremental improvements are often not enough, because fundamental changes in production and consumption systems are required in order to meet the needs and aspirations of a growing world population while using environmental resources in a sustainable manner (IHDP, 1999, p. xi). Solving the climate change problem, for example, demands new technologies that utilise renewable energy sources for energy and transport systems instead of the fossil fuels commonly used. They therefore require radical departures from existing technological and user practices. Figure 7.1 shows how various modes of radical innovation (in the upper left and

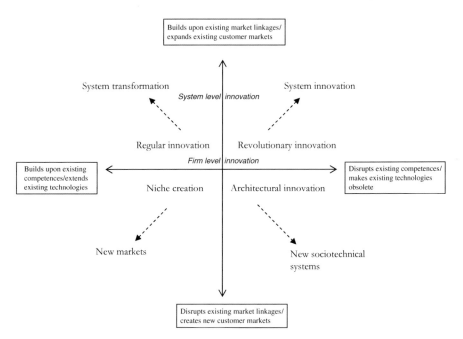

Fig. 7.1 Typology of firm level innovation and system level innovation (Hofman, 2005, p. 16, adapted from Abernathy and Clark (1985, p. 8)

lower quadrants) involve new technologies and products incorporating new competences and user functionalities while they make existing technologies obsolete and/or disrupt existing market linkages. In other words, radical innovations rely upon new sets of competences, technologies, functionalities and market linkages. As these radical innovations spread across firms and users, and are accompanied by changes in other technologies and in new user functionalities (e.g. forming a new technological system) they become part of a broader process of change at the system level.

Obviously, system change exceeds the capacity of individual firms. Firms that aim to develop innovative products and bring them to market face several constraints in accessing resources, information and knowledge from other actors in the regional innovation system. Especially in the case of more radical innovations that divert more substantially from existing patterns of production and consumption, various changes are required that go beyond the boundaries of firms, as is shown in Table 7.1.

Table 7.1 Radical innovation characteristics and type of changes required (Hofman, 2005, p. 48, extended from Abernathy and Clark (1985, p. 5))

Innovation aspect	Radical innovation characteristics	Type of change/actions required
1. Technology and production specific aspects		
– Design/embodiment of technology	Offers new design, radical departure from past embodiment	Find out what kind of design fits both technology and society
– Production system/organisation	Demands new system, procedures, organisation	Gain experience with production techniques and organisation
– Skills (labour, managerial, technical)	Destroys value of existing expertise	Re-train workforce, recruit new labour, built new expertise
– Material/supplier relations	Extensive material substitution; opening new relations with new vendors	Search for reliable and cheap materials, find reliable suppliers
– Capital equipment	Extensive replacement of existing capital with new types of equipment	Find, develop appropriate equipment and reliable equipment suppliers
– Knowledge and experience base	Establishes link to whole new scientific discipline, destroys value of existing knowledge base	Tap and find new sources for type of knowledge required, built new knowledge base
2. Market and customer specific aspects		
– Relationship with customer base	Attracts extensive new customer group, creates new market	Find out what the new market is for the innovation, what are appropriate niche markets

Table 7.1   (continued)

Innovation aspect	Radical innovation characteristics	Type of change/actions required
– Customer applications	Creates new set of applications, new set of customer needs	Customise product to potential application and user preferences
– Channels of distribution and service	Requires new channels of distribution, new service, aftermarket support	Modify and built up channels of distribution, service; develop competencies for maintenance
– Customer knowledge	Intensive new knowledge demand of customer, destroys value of customer experience	Set up pilots to analyse user behaviour to product/technology, develop means for educating users
– Modes of customer communication	Totally new modes of communication required	Develop appropriate modes of communication

Particularly SMEs face barriers ranging from more practical (lack of funds and time) to more strategic such as difficulty to access and appropriate information and knowledge, problems to generate market demand for their innovations, lack of insight in relevant regulations and polices, and lacking appropriate network partners (e.g. Hillary, 2000). Generally speaking, they lack the capability:

– to relate to or even influence the external environment and especially the regulatory context;
– to define a niche that their innovation could create or relate to;
– to attract venture capital.

Furthermore there is the technological challenge where often progress in a key technology needs to be accompanied by improvements in various complementary technologies. This implies that a network of actors is involved in an innovation process and this demands a level of coordination and management that can be difficult for SMEs to realize.

Intermediary organizations may provide various functions to bring in the resources and capabilities that SMEs lack, to facilitate the required changes as identified in Table 7.1 and to overcome barriers in the innovation process. Howells (2006, p. 720) identifies key functions such as:

– Foresight and diagnostics (e.g. technology roadmapping and needs articulation)
– Scanning and information processing
– Knowledge processing and combination/recombination
– Gatekeeping and brokering (combining knowledge of different partners)
– Testing and validation (early lab trials, pilots of innovations)

- Accreditation and standards (developing technical and industrial standards)
- Regulation (anticipating and influencing regulation)
- Protecting the results (intellectual property management)
- Commercialisation (finding lead users, develop marketing strategy)
- Evaluation of outcomes (evaluation and improvement of product performance)

In the remainder of this paper we focus on the question whether regional support structures fulfill the needs of firms. Ten cases of sustainable innovation will be analysed with regard to their support needs and the extent to which these were provided by regional support structures.

7.3 Functions of Regional Innovations Systems and Support Structures

With the advent of the systems of innovation approach the focus in regional development has increasingly shifted towards understanding the way processes of interactive learning can be stimulated through regional support structures. A range of actors have been identified (Todtling & Tripl, 2005) that play a role in regional innovation systems, see Fig. 7.2. According to these authors, the key for successful regional innovation systems is to create effective linkages between the different groups in the system, represented by the black arrows in the figure.

Building alliances is a key element in creating promising conditions for sustainable innovations as many of the needs we identified are related to the relationship the firm has with its commercial and regulatory environment. Smits and Kuhlmann (2004) formulate a number of functions at the level of innovation systems: (1) management of interfaces; (2) providing platforms for learning and experimenting; (3) providing an infrastructure for strategic intelligence; (4) stimulating demand articulation, strategy and vision development. Existing policy instruments only fulfil part of the systemic functions, and further development of systemic instruments is called for. This especially includes strengthening of the intermediary infrastructure comprising of institutions, mechanisms and organisations aimed at improving the interface and exchange of knowledge between the supply side and demand side (Smits & Kuhlmann, 2004, p. 16). Different types of of intermediary organisations have emerged in order to fulfil specific roles within innovation processes. These range from organisations that support knowledge diffusion and technology transfer to firms, organisations that provide bridges between different various actors and networks, organisations that support the search for funding, and organisations that support project management.

In an analysis of 10 sustainable innovation processes we will assess what the type of needs of firms were, and whether they were appropriately matched by organisation providing the functions as proposed by Smits and Kuhlmann (2004) and Howells (2006).

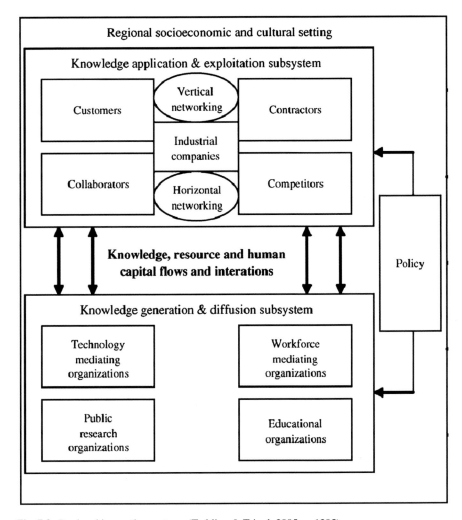

Fig. 7.2 Regional innovation systems (Todtling & Trippl, 2005, p. 1205)

7.4 Analysis of Sustainable Innovations

Ten sustainable innovations are analysed in detail to understand the support needs within the process and the support functions that were provided. The analysis is based upon a range of projects in which the authors were involved (Bonnick, Mora, Quiblat, Jiang, & Zheng, 2008; Hofman, 2005, 2006; Ruud, Lafferty, Marstrander, & Mosvold Larsen, 2007). Table 7.2 gives descriptions of the innovations included in the analysis. The analysis will be done in three blocks: first three larger Dutch companies, then two Norwegian companies, then five SMEs in

Table 7.2 Overview of the innovations included in the analysis

Innovation	Firm	Description
1. Biomass-fired electricity plant	Large energy company, the Netherlands	The biomass power plant has several innovative features, with some of them of a more incremental and some a more radical nature. Generating electricity by combustion of wood is clearly a proven technology. The plant is considered as innovative, because of its capacity and its efficiency of about 30% in the production of 24 Mwe through high temperature and steam pressures. More radical however are the logistics for the biomass input – 250,000 tonnes of annual wood input from various sources – and the fact that the product is marketed as 'green electricity', exploiting consumer appeal for its green image.
2. Phosphate product from process water	Large dairy company, the Netherlands	The innovation under study concerns the application of membrane technology for removing the relatively pure phosphate from the process water and is developed mainly due to regulatory pressures. The innovation process covers a 15 years period of trial and error. The side product that is produced turns out to be valuable and is exported. Installation of the membrane technology was possible without major changes in key production processes, but bringing the side-product to the market required the built up of new customer channels.
3. Water-based inks for printing	Large printing company, the Netherlands	By switching to water-based inks, the printing company will be able to ban all VOC's from its production process. The innovation is incremental in the sense that it does not change the product and the way it is marketed. The innovation is radical in the sense that it does disrupt the existing technological relationships in the chain from supply to production to re-use of printed paper. In the product chain, the water-based inks cause certain problems as the regular process of de-inking, necessary for the re-use of used paper, is frustrated by the use of this type of ink.
4. Wafer production for photovoltaic panels	Start-up firm, Norway	Start-up firm for the production of multicrystalline wafers for solar panels. Since the first production line for multicrystalline wafers was opened in 1997, the firm has evolved into one of the leading manufacturers of wafers in the world. The enterprise was successful because it has mastered technological and marketing challenges of this radical innovation and emphasized improvements in its production processes innovations that make economies of scale possible.

Table 7.2 (continued)

Innovation	Firm	Description
5. Heating and cooling with CO_2 as medium	Small start-up company, spin-off from large energy firm, Norway	This innovation involves the development of heat exchangers with CO_2 as a medium. It started in academic circles in Norway, where a professor in refrigeration engineering invented ways to effectively use carbon dioxide as a medium, triggered by the phasing out of CFCs as cooling medium in the Montreal protocol. The research was financially supported by a large energy firm who was interested in an application of the technology for air-conditioning systems in cars, mobile air conditioning (MAC). However, the first market application did not occur in cars but in households. The technology was applied in a small, efficient water heater a license agreement facilitated the use of heat pumps for tap water in Japanese homes.
6. Worm reactor	Start-up firm (joint-venture entrepreneur and water board) north Netherlands	A worm reactor can reduce sludge in water treatment plants by an estimated 50%. The entrepreneur developed up the idea from bio membrane projects at research institutes. Lab tests and pilot tests at a water board have taken place, key ideas are patented (reactor and worm specifics). Further testing at market scale is taking place to facilitate use by other water boards. The innovation is radical in the sense that it develops new techniques for sludge treatment but builds upon existing market linkages.
7. Electric bicycle	Small entrepreneurial network, north Netherlands	Drymer is a roofed, three-wheeled human-powered vehicle with an electric engine that doubles human pedalling power. It is expected to provide a mobility alternative to the car for small and medium distances.
8. Membrane reactor for wastewater	SME engineering firm, north Netherlands	The membrane reactor system is a small compact unit for waste water treatment and the water after treatment can be directly used in the household or other areas.
9. Pure plant oil as fuels	Small enterprise, north Netherlands	Pure plant oil (PPO) is derived from rapeseed for use as alternative fuel for vehicles. The project involves rapeseed farming, extraction, and conversion of vehicle engines to run on PPO. The project started in 2002 and has converted several hundred vehicles to run on PPO, but is now on indefinite status following changes in Dutch policy governing biofuels.
10. Virtual homecare	Small telecom firm, north Netherlands	The visual Care at Home is a public private partnership set by a small telecom firm in collaboration with the local health care organization, which serves approximately 50.000 people in Friesland by visiting them at their homes and providing assistance. The project is focused on e-health and its scope is to provide an alternative digital service to the traditional health care service by using audio-visual and IP technology.

the North of Netherlands. The three questions posed in the introduction will be discussed in separate sections.

The 10 cases differ with regard to key characteristics of the innovation. However, most innovations involve both changes in technology and market linkages as discussed in Table 7.1, with the exception of cases 3 and 6 that involve mainly technological changes while building upon existing market linkages (e.g. revolutionary innovations in the typology provided). Key differences occur with regard to the extent that the innovation disrupts existing technological set-ups and market linkages. In the first case, for example, proven technology is used for biomass combustion, but the main challenges remain the supply and logistics of biomass, and the build up of trustworthiness and a customer base for the distinct 'green electricity' product. The two other energy innovation cases (4 and 9) divert further from existing technological set-ups and market linkages. They call technology-wise for changes in production, capital equipment, materials and skills and knowledge, while also demanding development of new markets and customer relations, particularly for case 4. The cases of innovations that develop new water treatment techniques, cases 2, 6 and 8, are less far-reaching in the sense that technology-wise they do not imply broader changes in technological systems, while for case 2 and especially 8 the disruption of market linkages is significant. The proposed new mobility concept in case 7 is far-reaching both in terms of technology and customer aspects, while case 10 brings a new service into existing customer relationships and involves a new concept and design for health care. In the following sections we will answer the three questions posed in the introduction.

7.4.1 What Are the Suppport Needs of Firms in Sustainable Innovation Processes?

An overview of the role of support actors in the innovation processes is provided in Table 7.3. We first look at the three larger firms involved in sustainable innovation in the Netherlands (cases 1–3). For all three cases the innovating firms were able to tap knowledge from actors outside the firm and established effective networks for technology development and within their supply chains. For cases 1 and 2 interactions with regional regulatory bodies played an important part in the innovative processes and the companies successfully negotiated leeway to move the innovations forward. For case 3 the technology network was highly international with firms along its supply chain. Further diffusion of the innovation was however hampered by the resistance within the paper recycling chain and by lack of demand. Establishing markets for the new product innovations was also an important challenge in cases 1 and 2 (green electricity and the phosphate by-product). The two firms were successful in bringing the new product to the market, and building new customer channels, as they could build upon their existing marketing strengths and were able to develop networks with new partners that brought in key resources, e.g. marketing of green electricity and improving trustworthiness of the product was strengthened by a partnership with a leading environmental NGO (Hofman,

Table 7.3 The role of regional support actors in the innovation processes

Innovation	Role of regional support structures
Biomass-fired electricity plant (Hofman, 2002, 2005)	In the implementation of the innovation the company had to overcome several impediments related to the discussion on the character of the input (waste or biofuel), the actual various sources for the biomass, the emission standards, and the energetic efficiency of the power plant. The company was able to progress through various rounds of discussions and negotiations because it is a relatively powerful player in the Dutch electricity sector and has good established contacts both at the provincial and national level.
Phosphate product from process water (Hofman, 2000, 2005)	The network configuration for this case was initially a policy oriented network, constituted by the case company, its two subsidiaries, and the legal authorities for these subsidiaries, a regional waterboard and a national water directorate. The interplay between the case company and the waterboard lead to a search for a technical solution for the phosphate emissions problem. As a result the network was enlarged with several R&D organisations, which caused a shift towards a R&D oriented network. A crucial fact has been that the waterboard didn't urge for a quick solution, the application of end-of-pipe technology, but encouraged the company to search for a process integrated solution. There was also a personal factor involved, the particular officer who offered the company this leeway. The network has been an important factor, although it concerned an ad hoc network, that has developed around the specific innovation process, and that was not likely to remain afterwards. As a result the company was able to transform the phosphate rich waste water stream into a valuable side-product.
Water-based inks for printing (Hofman, 2001, 2005)	The network configuration in this case was predominantly R&D oriented. The firm had contacts with consultancies and R&D organisations with involvement of producers of inks, paper and printing machinery. Notably the inks producer has made a substantial effort for the innovation project also based on the long relationship it already had with the printing firm. Policy makers and legal authorities played a background role. Apart from the company's standing relationships with the inks and paper producers, it concerned an ad-hoc production chain oriented network, that emerged while the innovation project was performed. This network was also highly international due to the role of the producers of equipment, paper and inks, all foreign companies.
Wafer production for photovoltaic panels (Ruud et al., 2007)	The role of the Norwegian Industrial and Regional Development Fund (SND) was an important factor during the initial phase. As various sales contracts were established, SND agreed to cover 25% of the total investment. This enabled expansion as further large-scale private investors became interested and participated. Public subsidy programmes for solar energy abroad – particularly in Japan and Germany – combined with the regional-local potential of taking over existing facilities and highly suitable workforces, laid the foundation for REC ScanWafers success.

Table 7.3 (continued)

Innovation	Role of regional support structures
Heating and cooling with CO_2 as medium (Ruud et al., 2007)	Key actors were to be found within an existing energy company and a research institute where the technology itself was developed. The energy company initiated specific R&D activities enabling worldwide patenting. Other R&D initiatives funded by the EU have strengthened the commercial efforts, but more stringent regulatory standards than the current EU MAC directive are needed for further dissemination. National public authorities in Norway or the USA have not played an active role to promote the introduction of the technology.
Worm reactor (Bonnick et al., 2008)	The potential for sludge reduction by the worms was coincidentally discovered in a biomembrane reactor project in a money meets ideas programme that links entrepreneurs to investors. Close relations with potential users (water boards) facilitated further development. Further funding was obtained by the applied research organization for water management.
Electric bicycle (Bonnick et al., 2008)	The project is a follow-up of an earlier project where a prototype was developed called MITKA under the lead of the research institute TNO. An improved concept Drymer was developed by a regional technology centre in collaboration with several firms, for the phase of market introduction a number of firms organized their own entrepreneurial network but had difficulty in establishing initial markets for the product.
Membrane reactor for water treatment (Bonnick et al., 2008)	Through the Fryslan Water Alliance, the firm found some actors they can partner with for developing the product. The Friesland government provided subsidy for developing the product, while the water board effectively operated as matchmaker between the firm and the University of Twente to allow for research and testing facilities. A demonstration project was organized in collaboration with a municipality, province and the water board.
Pure plant oil as fuels (Bonnick et al., 2008)	The firm was able to organize a network with a company for oil extraction, an union of rapeseed farmers, and a network of vehicle dealerships with engineers trained in engine conversion technology. Local government was supportive and helped to obtain first customers, national duty exemptions were granted but later withdrawn. The company had difficulty obtaining funds to develop the innovation.
Virtual homecare (Bonnick et al., 2008)	The initial pilot project was financed by the province, and initial users are satisfied with the project. The firm has difficulty to move from pilot to market introduction as the subsidy is not continued. Initially the company expected to have a monopoly for this type of service in the region, later it became clear that other organization could also deliver the services.

2002, 2008). In case 1 regional support played a role through local government as launching customer (Hofman, 2002). In case 3 the unwillingness of government organizations to move towards use of water-based printing for their printing materials was cited as one of the factors contributing to failure of the innovation (Hofman, 2001).

The company in case 4 was rather successful to move from a start-up to major player in the solar energy market. Regional support played an important role during its start-up phase as it received significant financial support to set up production facilities. The company has been able to establish itself and become successful based on a strong knowledge network related to technologies. Further it has created a strong supply network related to the various steps of manufacturing of silicon to multi-crystalline wafers for solar cells, modules and panels, enabled by the personal experiences, capacities, and networks of the founder of the company. Case 5 reflects international negotiations on technologies to mitigate climate change. With regard to the mobile air conditioning (MAC) application in cars the innovative process has mainly involved positioning of the technology through the formation of alliances and networks, and through positioning the technology as a realistic option for greenhouse gas abatement and CFC and HFC phasing out, with specific focus as CO_2 technology as an alternative for HFC-134a technology in MAC.[1] To reach the situation where CO_2 technology was considered as a serious option, a development period of around 15 years took place. In this period the company significantly invested in R&D to improve the concept, developed networks with various R&D centres and car makers, and collaborated in various research projects funded by industry and governments. The innovation process itself has been one of manoeuvring to create the right network partners and conditions, and to prevent other parties from hijacking the principles under the concept. Moreover it has been a highly politicized process, with the regulatory focus on global warming potential of various cooling media as one of the main issues (Hofman, 2006; Ruud et al., 2007).

Cases 6–10 all involve SMEs involved in sustainable innovations in the North of the Netherlands. The innovative process for case 6 has involved moving from the laboratory tests to pilots to market introduction and involved constant user-producer interaction as water boards offered their facilities. The entrepreneur received some subsidies but had difficulty obtaining capital from financial institutions. Also protecting intellectual property rights was problematic as the entrepreneur was interested to collaborate with a research institute but was hesitant to share the key ideas. Case 7 was relatively successful in building a working prototype, under project management of a technology centre. The technology centre pulled out of the project in the phase of market introduction, and the companies involved failed to find lead users for the product. The innovative process is characterized by a large number of actors collaborating (eight SMEs, several knowledge centres, receiving funding at regional, national and international level), but lacks a dedicated and committed leading company. The innovation in case 8 moved from idea to market introduction through close collaboration of the firm with water boards. Also a knowledge institute was involved with regard to testing. The product is intended for decentralized

---

[1] An important factor in the process was the formation of an EC-directive involving sharpened emission standards for MAC. A ban on HFC-134a is proposed by the EC directive for new cars by 2017, and a ban on new car models with HFC-134a by 2011, with a phasing out process before that. This implies that the prospects for alternative concepts for mobile air conditioning, such as CO_2 based technology are promising.

water treatment and main market opportunities exist in Eastern Europe. Demand has been stagnating because potential users lack funds, and because policies shifted from favouring decentralized systems to more large-scale centralized wastewater treatment systems. The innovation in case 9 was significantly triggered by national bio fuel policies and the 2003 EU directive on bio fuels. The company was able to build a network facilitated by local support and national fiscal incentives. The firms' main problem was to create stable financial conditions. Tax exemption was withdrawn as the firm had to source part of its plant oil from foreign sources. The firm had difficulty obtaining credit from commercial banks, and lacked a proper marketing strategy. In case 10 the innovating firm started the project as it saw the potential of e-health schemes and was supported by regional and national public funds. After an initial successful pilot project the firm has difficulty bringing the services further into the market as financial support was discontinued and it faced competition whereas it had earlier assumed it occupied a monopoly position.

7.4.2 How Do These Needs Differ Across Types of Firms and Innovations, e.g. SMEs vs Large Firms, and More Incremental vs Sustainable (Radical) Innovations?

Overall, the key needs that can be identified in the firms centre on the role of demand, finance and regulation. This particularly reflects the following support function identified by Howells (2006): needs articulation, anticipating and influencing regulation and finding lead users and developing a marketing strategy. While anticipating and influencing regulation is a key problematic for both larger and smaller firms, finding initial markets and lead users is particularly problematic for smaller firms. This suggests a significant divide between needs for regional support for SMEs and larger firms. Large, well established firms are better able to organise the innovative process and its networks. If specific needs for regional support occur they are often well place to gain access with regional partners and to negotiate with regional government bodies. The picture is much bleaker for SMEs. These generally have difficulty in getting access to finance, are faced with regulatory barriers and policy shifts they feel they cannot influence, and especially have difficulty in targeting and developing a market for their innovation. The cases also suggest that for more radical innovations (e.g. pure plant oil, cooling with CO_2 as medium, the electric bicycle) key problems are related to the regulatory environment as often policy changes need to facilitate the innovation, and to the development of new markets for their innovation. Effective regional support structures and intermediary organisations therefore can play a crucial role in facilitating SMEs in their innovative process. With regard to the technological and production aspects of the innovations, the cases make clear that developing the innovation is mostly a collaborative effort with different partners bringing in specific technological competences. While larger firms are able to build their own technological networks, regional support can be helpful in helping SMEs find the appropriate technological partners and also in organising the network.

7.4.3 *How Effective Are Regional Support Actors in Providing Functions in Sustainable Innovation Processes?*

Table 7.4 takes into account the various functions provided by intermediary organizations as proposed by Howells (2006) and assesses whether these were provided for the different innovations.

The functions that were needed most while not provided were foresight and diagnostics (function 1), regulation (function 7) and commercialisation (function 9), see Table 7.4. The table implicates two important conclusions. First, it indicates that especially larger firms were able to have their needs fulfilled, either as functions were provided in house or as support actors provided them. For SMEs key functions that they need are not provided by support structures whereas they do not possess the capability to deliver them in-house. Second, key support structure elements that are needed most by SMEs concern relating to their regulatory and commercial environment. The cases we looked at clearly showed the need for more insights into markets, niches and needs on the one hand (functions 1 and 9). Commercialisation strategies for gaining access to and attracting (venture) capital are also weak in SMEs. Also a clear need with regard to anticipating and influencing regulation was identified (function 7). Many SMEs also have difficulty with

Table 7.4   Assessment of provision of functions (case number in brackets, ih = in house provided)

Need/function	Needed and provided	Needed not provided
1. Foresight and diagnostics (e.g. technology roadmapping and needs articulation)	(1, ih)	(3) (5) (7) (9)
2. Scanning and information processing	(1, ih) (2)	
3. Knowledge processing and combination/recombination	(1, ih) (2, ih) (3, ih) (4, ih) (5, ih)	
4. Gatekeeping and brokering (combining knowledge of different partners)	(1, ih)	(6)
5. Testing and validation (early lab trials, pilots of innovations)	(6) (8) (10)	
6. Accreditation and standards (developing technical and industrial standards)	(5)	(9) (10)
7. Regulation (anticipating and influencing regulation)	(1) (2)	(3) (5) (9) (10)
8. Protecting the results (intellectual property management)	(4, ih) (5, ih)	(6)
9. Commercialisation (finding lead users, marketing)	(1, ih) (5, ih) (9)	(7) (8) (10)
10. Evaluation of outcomes (evaluation and improvement of product performance)	(2, ih) (4, ih) (6) (7) (8)	

interpreting and anticipating regulations and policy developments and are caught off guard as regulatory circumstances change (often in their disadvantage).

If we look at the more broader systems innovation functions proposed by Smits and Kuhlmann (2004), we conclude in Table 7.5 that especially for the areas of demand articulation, policy anticipation, market intelligence, strategy and vision development the five SMEs in the region of the North of the Netherlands could benefit from effective expansion of the functions (1), (3) and (4) in the regional innovation system. This holds especially for the innovations (7), (9) and (10). The innovations that involve new water treatment techniques are embedded in a water technology cluster that show several features of regional innovation system, involving diverse actors such as water boards, research institutes, firms and start-ups, and intermediary agencies and programs that promote interaction such as a water alliance, R&D and subsidy programmes, and schemes to connect ideas to venture capital. For the other innovations such features of regional innovation systems are found to be much weaker and further progress of these innovations could be facilitated by strengthening the specific functions of the regional innovation system.

Table 7.5 Functions of regional innovation systems

Regional innovation system function	Goal	Needed & provided	Needed not provided
1. Management of interfaces	More effective interfaces between policy and the market could allow SMEs to finetune their innovative processes with anticipated and/or uncertain policy developments.		(7) (8) (9) (10)
2. Providing platforms for learning and experimenting	User-producer interaction is often a crucial part in sustainable innovation. Especially in areas where certain technology clusters are missing, these platforms could support SMEs	(6) (8)	(7)
3. Providing an infrastructure for strategic intelligence	Strategic information (market development, policy development) is crucial for SMEs but goes beyond their capacity. Regional support could play a key role as provider.	(6) (8)	(7) (9) (10)
4. Stimulating demand articulation, strategy and vision development	Sustainable innovations often derive their potential from a connection to a sustainability vision. The cases illustrate that demand articulation and connection to regional visions is often deficient		(7) (9) (10)

7.5 Discussion and Conclusions

The paper analyses the emergence of sustainable innovations in a selected number of firms and addresses key explanatory factors that contribute to emergence and diffusion of the innovations. The focus is particularly on regional support structures that facilitated the innovation processes, and on gaps between the needs identified within firms' innovation processes and functions provided by support structures. Ten sustainable innovation processes are analysed to gain insight in the relationship between the nature of the innovation process, the type of needs for firms, and the type of functions provided in regional innovation systems.

We started the paper by addressing the type of changes needed for radical innovations. We used the typology developed by Abernathy and Clark (1985) and extended by Hofman (2005) and distinguished technology and production specific aspects on the one hand and market and customer relations specific aspects on the other. Successful innovations take as much place outside the company as inside. A firm needs to position its innovation in such a way that it fits customer needs (or even better: creates such needs) and complies with (future) regulatory demands. Especially SMEs lack the capacity and skills to relate to the external environment. They are focusing on the aspects they can oversee, mainly within their own direct surroundings. They are, in other words, more focused on company and technology specific aspects. Most of the barriers exist with regard to market and customer relations specific aspects.

It is concluded that especially for SMEs demand articulation remains a major barrier as users are often only involved when the innovation is ready to enter the market, while regional support functions in this respect are deficient. Moreover, SMEs have major difficulty interpreting and anticipating sustainability policies and regulations at local and national levels, leading to innovations that face major regulatory barriers or are unable to cope with policy changes. Finally, we identified access to capital as a serious barrier.

Can these barriers be overcome? What are the policy implications from our analysis? First of all, our conclusions show a need for regional support structures. Without these it is not to be expected that the contribution of SMEs to sustainable development will improve significantly. Key areas to focus on are threefold:

> One: provide a link to policy development and an analysis of relevant expected developments;
> Two: bridging the gap between the firm and (potential) users. This requires a creative function, thinking along with the firm in terms of markets, functions of products and users;
> Three: linking the firm to venture capital.

Key functions that could more effectively build regional support systems are therefore the stimulation of demand articulation and vision development, organizations that can provide strategic intelligence that SMEs can not obtain in house, and interfaces between policy and business that allow firms to better cope with policy uncertainty and anticipate policy developments.

Our analysis has shown that current support structures fail to deliver these functions sufficiently. The role of governments could be to make sure that these functions are better served. Improving current structures is to be preferred over establishing new ones as the latter will need ample time to get access to SMEs.

References

Abernathy, W. J., & Clark, K. B. (1985). Innovation: Mapping the winds of creative destruction. *Research Policy, 14,* 3–22.

Bonnick, A., Mora, D., Quiblat, N., Jiang, D., & Zheng, G. (2008). *Sustainable innovation and the role of regional actors.* Report prepared for the MSc of Energy and Environmental Management, Leeuwarden, CSTM-University of Twente.

De Bruijn, T., & Tukker, A. (Eds.). (2002). *Partnership and leadership; building alliances for a sustainable future.* Dordrecht: Kluwer Academic Publishers.

Dimitriadis, N., Simpson, M., & Andronikidis, A. (2005). Knowledge diffusion in localized economies of SMEs: The role of local supporting organizations. *Environment and Planning C, 23,* 799–814.

Freeman, C. (1987). *Technology policy and economic performance: Lessons from Japan.* London: Pinter.

Freeman, C. (1988). Japan: A new national system of innovation?. In G. Dosi, C. Freeman, G. Silverberg, & L. Soete (Eds.), *Technical change and economic theory* (pp. 330–348). London: Pinter.

Gerstlberger, W. (2004). Regional innovation systems and sustainability, selected examples of international discussion. *Technovation, 24,* 749–758.

Hartman, C. L., Hofman, P. S., & Stafford, E. R. (1999). Partnerships: A path to sustainability. *Business Strategy and the Environment, 8*(5), 255–266.

Hartman, C. L., Hofman, P. S., & Stafford, E. R. (2002). Environmental collaboration: Potential and limits. In T. J. N. M. De Bruijn & A. Tukker (Eds.), *Partnership and leadership – Building alliances for a sustainable future* (pp. 21–40). Dordrecht: Kluwer Academic Publishers.

Hillary, R. (Ed.). (2000). *Small and medium-sized enterprises and the environment.* Sheffield: Greenleaf Publishing.

Hofman, P. S. (2000). Innovation by Negotiation – *A Case Study of Innovation at The Dutch Dairy Company Borculo Domo Ingredients.* CSTM series no. 145, Enschede.

Hofman, P. S. (2001). *Innovation, negotiation and path dependencies in industry and policy – Environmental and technology policy induced innovation in the Netherlands.* Proceedings of the Ninth Greening of Industry Network Conference, Bangkok, 21–25 January, 2001.

Hofman, P. S. (2002). Becoming a first mover in green electricity supply: Corporate change driven by liberalisation and climate change. *Greener Management International, 39,* 99–108.

Hofman, P. S. (2005). *Innovation and institutional change, the transition to a sustainable electricity system.* PhD Thesis, University of Twente, Enschede.

Hofman, P. S. (2006). *Scenarios for industrial transformation: Perspectives on the CondEcol case studies.* ProSus Report no. 4/06, ProSus, University of Oslo.

Hofman, P. S. (2008). Governance for green electricity: Rule formation between market and hierarchy. *Energy and Environment, 19,* 803–817.

Howells, J. (2006). Intermediation and the role of intermediaries in innovation. *Research Policy, 35,* 715–728.

IHDP. (1999). *Industrial transformation science plan.* Bonn: International Human Dimensions Programme on Global Environmental Change, IHDP.

Lundvall, B.-A. (1988). Innovation as an interactive process. In G. Dosi, C. Freeman, G. Silverberg, & L. Soete (Eds.), *Technical change and economic theory.* London: Pinter.

Lundvall, B.-A. (Ed.). (1992). *National systems of innovation, towards a theory of innovation and interactive learning.* London: Pinter.

Lundvall, B.-A., Johnson, B., Andersen, E. S., & Dalum, B. (2002). National systems of production, innovation and competence building. *Research Policy, 31*, 213–231.

Morgan, K. (1997). The Learning Region: Institutions, Innovation and Regional Renewal. *Regional Studies, 31*(5), 491–503.

Ruud, A., Lafferty, W. M., Marstrander, R. O., & Mosvold Larsen, O. (2007). *Exploring the conditions for adapting existing techno-industrial processes to ecological premises: A summary of the CondEcol project* (ProSus report 2007/1). University of Oslo.

Smits, R., & Kuhlmann, S. (2004). The rise of systemic instruments in innovation policy. *International Journal of Foresight and Innovation Policy, 1*(1/2), 4–32.

Todtling, F., & Trippl, M. (2005). One size fits all? Towards a differentiated regional innovation policy approach. *Research Policy, 34*, 1203–1219.

Chapter 8
Disruption or Sustenance? An Institutional Analysis of the Sustainable Business Network in West Michigan

Deborah M. Steketee

Abstract Sustainable regional development is understood as a process characterized by continuing attention to lasting economic prosperity which supports healthy ecological and human communities within a defined spatial area. This preliminary research focuses on disruptive innovation network structures and institutional arrangements that are essential theoretical components for sustainability. The presumption is that by better understanding the characteristics of networks that support innovation, we may more effectively leverage sustainable business practices toward an effort to foster sustainable regional development. Network theory and institutional offers a way to illuminate some of the existing network nodes and institutions supporting sustainable business practices in West Michigan (USA) and identifying ways that learning associated with these practices might be harnessed in the service of sustainable regional development. Network centrality, reachability and connectivity are explored.

Keywords Disruptive innovation · Sustainable regional development · Network theory · Sustainability · Institutional analysis

8.1 Introduction

The lineage of recent innovation research distinguishes between game-changing 'disruptive' innovation and market-deepening 'sustaining' innovation (Christensen, 1997; Christensen & Raynor, 2003). Can this distinction help us understand how to catalyze sustainable regional development? Sustainable regional development is understood here as a process characterized by continuing attention to lasting economic prosperity which supports healthy ecological and human communities within

D.M. Steketee (✉)
Department of Sustainable Business, Aquinas College, Grand Rapids MI 49506, USA
e-mail: stekedeb@aquinas.edu

J. Sarkis (eds.), *Facilitating Sustainable Innovation through Collaboration,*
DOI 10.1007/978-90-481-3159-4_8, © Springer Science+Business Media B.V. 2010

a defined spatial area. In a context of dwindling natural resources and economic 'reset', innovation will be key to redesigning commerce to allow us to continually adapt to changing contexts. This research endeavors to illuminate network structures and institutional arrangements that will help couple innovation at a firm level with sustainability at a regional scale. The presumption is that networks which facilitate disruptive innovation toward sustainability in business may effectively embed a type of learning critical toward fostering sustainable regional development.

The heart of this descriptive case study rests on two related hypotheses. First, business-led disruptive innovation, rather than sustaining innovation, will be more effective in fostering transformative, sustainable regional development. Second, institutional arrangements supporting collaboration through networks will be central to regional success in this transformation as individual firm competency in innovation leverages sustainability as an organizing logic for regional development. The western region of Michigan offers a unique case for exploring these hypotheses. Although not immune to recent economic challenges in the U.S. economy, West Michigan is a region which appears to be successfully avoiding the downward spiral of other regions in the Great Lakes by replacing its 'rust belt' with a 'green' belt (West, 2008).

This chapter is organized into five parts. The following section describes innovation, distinguishing between disruptive and sustaining innovation, and exploring how innovation contributes to sustainability. The section also identifies recent contributions toward understanding regional innovation systems, offering insights on the linked set of nodes, connections and interactions that comprise effective collaboration networks. The third section further elaborates upon interactions, focusing on the essential role of institutions as guides and constraints for human interaction while pursuing a collective goal—in this case, sustainable regional development. This section sets the stage for the fourth part of the paper which includes a brief case study of the West Michigan region in North America. After a general overview of the West Michigan region, a brief institutional analysis of the interaction between actors and processes provides insights into disruptive innovation as a pathway for sustainable regional development in the United States. The final section offers some concluding remarks, including insights for policy-makers and areas for further research.

8.2 Conceptual Crossroads: The Intersection of Innovation, Sustainability, and Regional Development

This research weaves together several theoretical threads, including those related to innovation, sustainability, and regional development. The intent here is not to provide an extensive literature review, but rather to sufficiently situate these concepts in an effort to understand disruptive innovation and its link to sustainable regional development. Figure 8.1 is provided here as a way to envision key conceptual components.

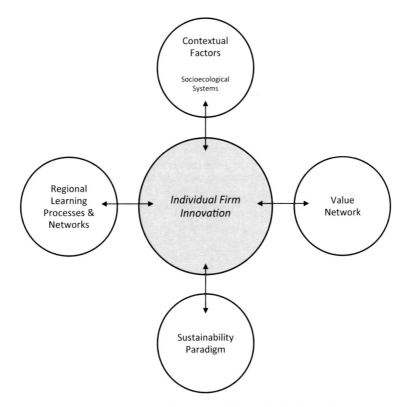

Fig. 8.1 Conceptual model of firm-level components for sustainable innovation

8.2.1 Understanding Innovation

Innovation may be understood as a process by which an individual or team of individuals reimagine and deploy new combinations of productive factors leading to new goods, services, processes (production and/or delivery), markets or ways to organize firms (Christensen & Raynor, 2003; OECD, 2005; Schumpeter, 1934). At its roots, innovation is a social endeavor (Meeus, Oerlemans, & van Dijck, 1999), playing a crucial role in an effective business strategy and situated within an increasingly dynamic and competitive global landscape (Prahalad & Krishnan, 2008). Innovation is interactive by its nature, involving a variety of actors both within and external to the firm. The diversity among ways of looking represents a key element of innovation—the ability to view new connections (Kelley & Littman, 2001).

Not all innovation is created equal, however. Building upon Joseph Schumpeter's process of 'creative destruction' (1934), recent work argues for innovation strategies based on circumstances (Christensen, 1997; Christensen & Raynor, 2003). Christensen and Raynor (2003) posit that 'disruptive circumstances' involve the challenge '...to commercialize a simpler, more convenient product that sells for less money' (p. 32). In other words, disruptive innovations have the potential to change everything as they meet new customer value propositions (Johnson, Christensen, & Kagermann, 2008). Disruptive innovation occurs among firms with more of a forward-looking approach to innovation, attempting to meet customer needs in unexpected ways.

On the other hand, 'sustaining[1] circumstances' involve 'a race that entails making better products that can be sold for more money to attractive customers' (Christensen & Raynor, 2003, p. 32). Sustaining innovations rely upon a customer's willingness to accept an 'improved' version of what they already have. Relying on retrospection, sustaining innovation builds incrementally upon successes of the past, giving incumbent firms an advantage (Christensen, 1997).

An organization's structure and its learning pathways have been identified as key to successful innovation of both types. As noted by Christensen and Raynor (2003), 'the organization's structure and the way its groups learn to work together can then effect the way it can and cannot design new products' (p. 34). Innovation also takes place within a 'value network'—described as 'the context within which a firm identifies and responds to customer needs, solves problems, procures inputs, reacts to competitors, and strives for profit...' (Christensen & Raynor, 2003, p. 36). The relationships within this value network are increasingly supporting a process of 'co-creation' (Prahalad & Krishnan, 2008) in which innovation becomes an open and iterative process between business, its customers and other stakeholders.

Looking through a conceptual lens of innovation, the rise of sustainability as a business imperative may signal an adaptation to disruptive circumstances— circumstances demanding that businesses respond with game-changing products, services, processes and/or business models and prepare to operate in a new value network. The end of 'cheap' raw materials and 'cheap' energy as well as the demands of an expanding and more complicated universe of stakeholder relationships lead away from business-as-usual toward new business models and strategies. Some businesses are shifting toward fundamentally different groundings, connecting to a paradigm focused on a redesign of commerce as an eco-effective (McDonough & Braungart, 2002) system which generates financial profit as it supports restoration of ecological systems and enriches the human condition. As transformative change, a paradigm of sustainability offers game-changing pathways for commerce.

While some businesses embark on a pathway toward transformation, others intentionally choose an incremental approach to sustainability. This latter approach, grounded in eco-efficiency (Esty & Winston, 2006), emphasizes 'sustaining' circumstances which advocate innovation that supports incremental change, particularly through reductions in environmental impact, rather than a rethinking of economic assumptions and values. The next section further explores these departures along the sustainability pathway.

---

[1]In fact, Christensen's use of the term 'sustaining' also refers to the continuing viability of a firm.

8.2.2 *Muddling Toward Sustainability*

In conventional business lexicon, sustainability previously referred to a firm's ability to last over time (Porter, 1985). This prior meaning dovetails with Christensen's idea of sustaining innovation. However, sustainability in today's business context typically signifies an increasing attention to the 'triple bottom line'—a focus on an integrated set of business practices which contribute to restoring ecological systems, fostering social equity and community well-being as well as assuring financial prosperity (Elkington, 1997; Hawken, 2007; Hawken, Lovins, & Lovins, 1999; McDonough & Braungart, 2002; Rainey, 2006). Pursuit toward a triple bottom line occurs within a broader effort to achieve sustainability—understood here as a condition where natural and social systems effectively align (Berkes & Folke, 1998), continuing indefinitely.

The term 'sustainability' has experienced tremendous popularity in recent years but also a lack of conceptual clarity (Faber, Jorna, & Van Engelen, 2005). The abundant rhetoric surrounding sustainability serves both as the concept's strength and weakness. The conceptual fuzziness of sustainability has allowed a wide swath of groups and individuals to embrace it and remake it for themselves (Hawken, 2007). However, in some ways, this conceptual noise also makes it difficult to promote collaborative strategies to bring a 'new industrial revolution' (McDonough & Braungart, 2002) to the factory floor, to communities and to regions around the globe. Practitioners who might choose to use sustainability's guiding principles for economic development (Gibbs, 2000) or business strategy (Van Bakel, Loorbach, Whiteman, & Rotmans, 2007) also tend to try to develop more generalized measures of sustainability—a 'one-size-fits all' reporting blueprint, which belies the complexity of sustainability and what will be required to govern aligned socio-ecological systems (Ostrom, 1995).

Despite this conceptual muddle, innovation is seen as critical in the business sector's pursuit of sustainability in its new guise (Council on Competitiveness, 2008; World Business Council for Sustainable Development, 2007). Suffice it to say that at the very least, the intersection of sustainability and innovation requires a rethinking of conventional, market-driven goals (Hall & Vredenburg, 2003). Sustainability, then, as a significant paradigm shift, represents a set of disruptive circumstances for businesses and the regions in which they exist.

Just as the tools of personal computing and communication technology have led to a recasting of our economy and ways of life, innovation fostered within the paradigm of sustainability has the potential to change the logic of the existing regional development system through learning that is based in experiences attempting to fundamentally redesign products, processes, management structures and business models. It is argued here that the networks and institutions required to support innovation—and in particular, disruptive innovation—embrace many of the essential characteristics which are key to effective pursuit of sustainable regional development. The next section provides an overview of the significance of networks for innovation and explores the question of whether and how disruptive innovation at the firm level can scale up to effect change toward sustainable development at a regional scale.

8.2.3 Scaling Up Through Networks

It has been argued that knowledge transfer is facilitated through social capital which adheres in collaborative networks (Inkpen & Tsang, 2005). Networks are critical to the innovation process. Proximity has been identified as an important component of innovation and the ability to transfer knowledge is linked to human interaction through networks (Broekel & Meder, 2008; Miguélez, Moreno, & Artis, 2008).

At a regional scale, networks exist in what has been described as regional innovation systems (RIS). Andersson and Karlsson (2004) capture three important points in understanding an RIS. First, they provide a useful definition of regions as 'a territory in which the interaction between the market actors and flows of goods and services create a regional economic system whose borders are determined by the point at which the magnitude of these interactions and flows change from one direction to another' (p. 7). This definition stresses the point that regions exist in space, and not simply place.

Second, systems consist not only of nodes, but also processes and interactions. Individual firms may be linked together through regional processes and networks. Through these processes and networks, benefits accrue not only to the individual firm but also to regional conditions. Benefits are similar at both the firm and regional level, including reduced transaction costs to exchange information, resolve disputes, transfer complex knowledge, access a wide stock of knowledge, and develop cutting-edge connections (Powell & Grodal, 2005).

Third, Andersson and Karlsson stress the flexible boundaries which characterize today's regions. It has been argued that while understanding the structural framework of regional innovation systems is important, it is critical to focus on the processes which take place within that structure (De la Mothe & Paquet, 1998). This argument stresses the need to consider regions in a more holistic view—defining 'place' by relationships as well as territory (Amin, 2004). Systems of relationships—understood as 'social capital across economic space' (Murphy, 2006)—have been identified as critical to innovation (Chaminade & Vang, 2006) because of their ability to foster the learning required for the increasingly collaborative endeavor of innovation (MacKinnon, Cumbers, & Chapman, 2002; Powell & Grodal, 2005).

Networks facilitate a collaborative learning process which may lead to positive or negative outcomes (Broekel & Meder, 2008; Morgan, 1997). One recent study examining regions in Spain, identified social networks, norms and trust (constituting 'social capital' in this research) as having a direct, positive impact on innovation outcomes (measured as patent applications) (Miguélez et al., 2008). However, it has also been found that too much cooperation limited only to intra-regional partners can actually have a negative outcome for innovation—what has been described as 'regional lock-in' (Broekel & Meder, 2008). In this situation, parochialism impedes the flow of new ideas into the area, blocking insights which feed creativity. Other patterns of interaction may lead to 'regional lock-out,' where potential partners may share the same place (region) but these intra-regional partners may not belong to the 'right network' within that physical location (Broekel & Meder, 2008; Castells, 2000).

In addition to their capacity to support knowledge transfer, collaborative networks serve as critical governance structures. Network structure appears to have significant advantages when dealing with complex and uncertain policy environments where '...loose, decentralized, dense *networks* [emphasis added] of institutions and actors that are able to quickly relay information and provide sufficient redundancies in the performance of functions so that the elimination or inactivity by one institution does not jeopardize the entire network' (Haas, 2004, p. 7). In a similar vein, ' multi-level' or 'polycentric'[2] governance has been shown to be a more effective way to structure social-ecological interactions in order to provide 'adaptive potential', consisting of attributes which allow the social-ecological system to meet expected and, even more to the point, unexpected challenges (Anderies, Janssen, & Ostrom, 2004; Gunderson & Holling, 2002; Ostrom & Janssen, 2002).

What links these conceptual ideas together is the notion that networks and the institutions in play within that network will shape the possibilities for fostering sustainable regional development. Sustainable development carries with it new dynamics in decision-making, expanding choices and opportunities for interaction, as well new time horizons for actors within an innovation system. If we accept the premise that both innovation and development are social processes, then there is reason to better understand social choices and their significance in terms of the desired outcome of a sustainable region. The next section attempts to sift through this complexity, providing observations about the role of institutions in linking innovation with sustainable regional development.

8.3 Understanding the Role of Institutions in Fostering Innovation for Sustainability

Factors relating to sustainable innovation are emerging in the literature (See for example, Borup, 2005; Green & Randles, 2006; Könnölä & Unruh, 2007; MacKinnon et al., 2002; Williams & Markusson, 2002), supporting earlier research stressing the significance of interactions between actors in the innovation process (Lundvall, 1992; Meeus et al., 1999). Embedded within these factors are sets of institutions. North (1990, p. 3) defines institutions as: 'the rules of the game in a society or, more formally, are the humanly devised constraints that shape human

---

[2]Vincent Ostrom's definition of a polycentric order as noted in Ostrom and Janssen (2002) is particularly useful. He described this order as one 'where many elements are capable of making mutual adjustments for ordering their relationships with one another within a general system of rules where each element acts with independence of other elements' (V. Ostrom, 1999, p. 57).

interaction... [reducing] uncertainty by providing a structure to everyday life'. These rules guide us by 'influencing the availability of information and resources, by shaping incentives, and by establishing the basic rules of social transactions' (Nicholson, 1993, p. 4).

Institutions allow factors to increase or decrease in importance. Context shapes the nature of and duration of interplay between factors; as well, the outcome of that interplay will also be shaped by institutions. For example, recent research indicates the innovations in clean production technologies result from deployment of management tools (cost savings, general management systems, environmental management systems, et al.), while regulatory requirements and policies tend to lead toward end-of-the pipe technologies (Frondel, Horbach, & Rennings, 2007).

Sustainable development requires institutions which will encourage individuals to see their individual self-interest aligned with the common good. Institutions are central to collective action strategies, since they constrain or expand possible choices in decision-making. Some institutions may lead to the same outcome, but involve different strategies over different time horizons. For example, implementation of laws regulating toxic waste may guide a local business professional to transition quickly from utilizing highly profitable synthetic chemicals to bio-based products. Another businessperson might be encouraged to make the transition well in advance of a law because faith-based community norms support stewardship of natural systems even if it generates a less profitable financial outcome over the short term.

Institutional analysis allows for closer examination of institutions in given situations and settings. For this selected case, the Institutional Analysis and Design (IAD) Framework[3] provides useful scaffolding for understanding institutional arrangements supporting sustainable regional development, taking context into account (Ostrom, 2005). The analyst may use appropriate theories within the framework, based upon the particular research question at hand. In this instance, network theory related to innovation is an appropriate anchor for the IAD framework.

The focal point of institutional analysis is the action arena (Ostrom, 2005) (See Fig. 8.2). Within an action arena, participants interact in various situations shaped by context, which is a collection of exogenous variables. The patterns of interactions are affected and effected by their context, leading to certain outcomes. These outcomes then may be evaluated according to chosen criteria such as economic efficiency, sustainability, economic productivity, social equity and others. In the IAD framework, evaluative measures may be applied to outcomes or the process of achieving outcomes (Gibson, Andersson, Ostrom, & Shivakumar, 2005; Ostrom, 2005, pp. 66–67).

---

[3]The IAD framework was originally developed by Elinor and Vincent Ostrom, along with numerous colleagues associated with the Indiana University Workshop on Political Theory and Policy Analysis. It continues to be refined as a result of its extensive use in helping to address a wide range of research questions.

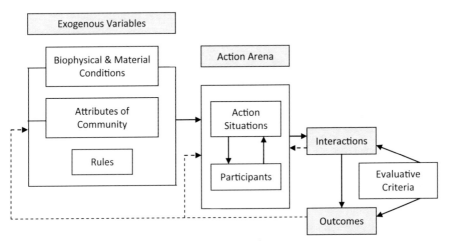

Fig. 8.2 Institutional analysis and design framework (*Source:* Ostrom, 2005, p. 15)

Outcomes loop back to the action arena, changing actors and the situation. Over time, feedback from outcomes may also affect context (Ostrom, 2005). Each of the general elements of the IAD framework may be unpacked, offering a more detailed understanding of the IAD framework.

Exogenous variables include the characteristics of a biophysical and material world, a cultural world and the rules used by actors. These 'context clusters' (Steketee, 2006) order actors' relationships in particular situations (Ostrom, 2005; Ostrom, Gardner, & Walker, 1994).

The first context cluster relates to *rules*—both formal (laws, regulations, et al.) and informal (how things are done). Ostrom defines rules as: 'the set of instructions for creating an action situation in a particular environment' (2005, p. 17). Variables in the second context cluster, *biophysical and material conditions*, attempt to reflect the physical world in which action takes place, affecting the 'physical possibility of actions, the producibility of outcomes and the knowledge of actors' (Ostrom et al., 1994). The final set of variables includes the *attributes of community* such as: 'the values of behavior generally accepted in a community; the level of common understanding that potential participants share (or do not share) about the particular types of action arenas; the extent of homogeneity in the preferences of those living in a community; the size and composition of the relevant community; and the extent of inequality of basic assets among those affected' (Ostrom, 2005, pp. 26–27).

The *action arena* is comprised of an *action situation* and *actors*. An actor does not necessarily mean a person acting singly; it may involve a group of individuals 'functioning as a corporate actor' (Gibson et al., 2005, p. 27). An action situation occurs when two or more actors interact to jointly produce an outcome (Ostrom, 2005).

An institutional analyst uses the action situation in order to 'isolate the imme-diate structure affecting a particular process in order to explain regularities in human action and results' (Gibson et al., 2005, p. 31). These outcomes can then be evaluated according to selected criteria.

8.4 A Brief Institutional Analysis of the West Michigan Region (Michigan, USA)

The central question for this brief institutional analysis is: What are the institutional arrangements which facilitate learning for sustainable regional development in the West Michigan region? As noted above, learning, or the process of tacit knowledge transfer, is seen as essential in successful innovation and even more important as part of the sustainable innovation process (Cohen & Levinthal, 1990; Hall & Vredenburg, 2003; Keane & Allisson, 1999; Williams & Markusson, 2002). This analysis takes a closer look at the existing networks which support sustainable business innovation and increase opportunities for social learning in order to transfer that knowledge to effect sustainable regional development. The evaluative criteria in this case focuses on the openness of these networks to both intra-regional and inter-regional learning to support sustainable business innovation. Specifically, the significance of the networks' level of connectivity, their level of centrality and reachability (Janssen et al., 2006) can be brought to bear in understanding these activities in West Michigan region.

In examining the connectivity tying together a network related to sustainable business, one may see a great density in the connections (according to Janssen et al., the number of links divided by the maximum possible number of links) as well as high level of reachability (the extent to which all of the nodes in the network are accessible to each other). This type of network structure and the norms embedded within that structure are seen to benefit learning. However, the caution is noted that if the density of social links is too high, it could hinder innovation, disrupting the balance 'between learning from others and room for individual innovation' (p. 5).

In looking at the level of the centrality of a network ('distribution of the links among the nodes in the network as well as their structural importance', p. 6), Janssen et al. argue that the distribution of information may be reduced if the centrality is high. Diversity in a network can be highly valued for the increased flow of information. A central node or actor insisting on centrality (e.g., 'all things flow through here') diminishes the flow of information and negates the benefits of network diversity.

8.4.1 Methodology

Primary research for this abbreviated institutional analysis utilized data collected through a process to identify and describe the 'nodes' and linkages in the West Michigan region. Data were gathered in multiple ways including phone surveys, cognitive mapping, web site research and key informant inquiries in order to better chart West Michigan's sustainability landscape. The information was initially collected as part of an on-going 'asset mapping project' being undertaken by the Center for Sustainability at Aquinas College[4] to identify the various nodes—both organizations and individuals—who were key players in regional and local sustainability efforts. Additionally, information was gathered at a summit of the Community Sustainability Partnership (see note 6), conducted in October 2007. During a facilitated session at the summit, 65 individuals participated in groups of four, five or six participants. Each group was asked to draw cognitive maps of the region's sustainability landscape identifying key nodes (both organizations and individuals) as well as connections between these nodes. There were no guidelines as to how this information was to be organized; each group was able to bring their own meaning to sustainability as they constructed their drawings. However, after a first round of drawing, these cognitive network maps were posted around the room and participants were asked to view each other's drawings. They were then allowed to return to their original groups and modify or augment their conceptions. Information from these drawings was reviewed to help identify key nodes of activity as well as linkages and resources in the region, particularly those related to sustainable business.

Following the mapping, 38 15–20 min phone interviews were conducted between January 2008–May 2008 with individuals whose organizations were endorsing partners of the Community Sustainability Partnership (CSP) and for whom contact information was provided by CSP administrators. In addition to identifying questions about their organization, interviewees were asked four short questions in order to identify current sustainability-related projects their organization was undertaking either alone or in partnership with other organizations or individuals. They were also asked to list other organizations or individuals they were aware of who were involved in sustainability-related efforts either on a local or regional basis. Responses were noted in written form on single sheets and then entered into a data base categorizing participants by sector of involvement. Sectors included environmental, government, faith-based, social services, education and business organizations. Additional information was gathered through informal conversations with key informants selected purposefully as a result of the author's participation in various sustainability efforts in the region. Secondary data was also reviewed, included various technical reports, newspaper articles, and on-line materials relating to regional sustainability.

For the purposes of this study, the West Michigan region is defined spatially as an eight-county area in Michigan, including the counties of Allegan, Barry, Ionia, Kent,

---

[4]The Center for Sustainability at Aquinas College serves as an outreach and research center in support of sustainable business and community sustainability. Some support for this effort was provided through the Community Sustainability Partnership, which was established to provide assistance to the City of Grand Rapids and Grand Rapids Public Schools. Aquinas College was one of five founding partners of the CSP.

Fig. 8.3 Map of the eight-county west Michigan region (*Source:* West Michigan Strategic Alliance, www.wm-alliance.org)

Montcalm, Muskegon, Newaygo and Ottawa (See Fig. 8.3) Three main cities— Grand Rapids to the east, and Holland and Muskegon on the Lake Michigan shore, anchor what has been called a 'metro triplex' (West Michigan Strategic Alliance, 2002). In this particular region the area's spatial boundaries coincide relatively well with the functional boundaries of this socio-economic region.

8.4.2 Exogenous Variables

8.4.2.1 Biophysical and Material Conditions

West Michigan, the region of interest for this study, is located in the Great Lakes basin of the North American continent. One of its most distinguishing characteristics is that the region is part of the Great Lakes basin, which contains 20% of the world's surface freshwater. The West Michigan region, bounded to the west by Lake Michigan, has developed as a result of its proximity to fast flowing rivers, which supported a thriving timber industry early in its economic history. It is still a region known for its rich natural resource base, including parts of the largest freshwater dune system in the world, giving it the moniker, Michigan's 'West Coast.' However, the region is becoming increasingly urbanized, particularly near the shore of Lake Michigan, resulting in environmental impacts typically associated with urban development such as water quality degradation, increased air pollution, and congestion.

8.4.2.2 Attributes of Community

The population in the combined metropolitan statistical areas has increased by 6% between 2000–2007, growing from 1,083,174 to 1,330,384 individuals (The Right Place, 2008), with a median household income of $51,524. The 2007 ethnic landscape was predominantly white (86%), with smaller representation of other ethnic groups including African-American (6%) and Hispanic (7%) (The Right Place, 2008).

The economic landscape is a diverse one, although manufacturing still plays a predominant role. Known once as the 'Furniture Capital of the World,' (and more recently as the 'Office Furniture Capital of the World'), more than 2,100 manufacturers are located in the region, representing a diversity of manufacturing types (The Right Place and WMSA, 2007, p. 18–19). Approximately 21.3% of the workforce is employed in manufacturing (The Right Place, 2009). However, unemployment (not seasonally adjusted) in 2007 stood at 6.2%, which was above the national average of 4.6% but below the Michigan average of 7.2%. Unemployment in 2008 had increased to 7.3% regionally, while Michigan's preliminary annual average had risen to 8.4% (State of Michigan, 2009).

The Right Place, a regional economic development entity, reports that the manufacturing payroll represents 34.1% of the area's total payroll, while the manufacturing sector represents 8.2% of establishments with payroll (p. 15).

In 2005, the region rated 6th out of 125 studied regions in the World Knowledge Competitive Index (The Right Place, 2008), based upon a composite of regional economy outputs, human capital, financial and knowledge capital and knowledge sustainability metrics.

8.4.2.3 Rules

The formal rule structure within the region offers a tangle of political jurisdictions. There are 219 municipal (local) government entities within the region, and 77 school districts overlaid with a variety of state agency jurisdictions. A regional planning entity, the Grand Valley Metro Council, and the West Michigan Strategic Alliance, a non-governmental organization dedicated to fostering a high quality of life for the region, both focus on regional scale efforts. These latter two organizations have gone through extensive 'visioning' processes in order to create a roadmap for the region's overall development, although with varying degrees of follow-through and outcomes.

There are also strong cultural norms of collaboration within the region. Some of these norms may be connected to deep and broad faith-based structures in the region, providing a higher than expected level of associational involvement and faith-based engagement in Grand Rapids, the largest city in the region (Grand Rapids Community Foundation, 2001). West Michigan's strong heritage of problem-solving and planning for the future has been credited by some as the reason for its on-going prosperity in the midst of Michigan's economic crisis (Zwaniecki, 2009).

8.4.3 Action Arena/Interactions

The action arena of interest for this research is the region's collection of efforts to develop the West Michigan region as a leader in sustainable business. A growing and interwoven set of networks related to sustainable business practices may potentially help move the region forward (see Fig. 8.4). A description of some of the network relationships and activities are provided here. Throughout this network, momentum for regional economic development appears to be anchored to an emerging strategy in which sustainability plays a central role. This strategy has been driven in large part by private sector initiative, and far in advance of the national business sector's embrace of sustainability.

Early corporate leadership was exemplified by office furniture makers Herman Miller, Inc. and Steelcase, Inc. (founded as private companies but both now public) as well as other privately-held, and lesser-known companies. For example, Herman Miller's founder D. J. DePree included environmental stewardship in a corporate

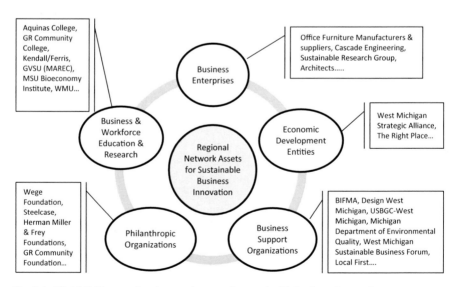

Fig. 8.4 West Michigan regional network assets for sustainable business innovation

values statement prepared in the 1950s (Rossi, Charron, Wing, & Ewell, 2006). The active engagement of succeeding generations of CEOs as well as strong advocacy and practices of environmental health and safety professionals (a number of whom have new titles with sustainability in the name) have led toward strong and swift attention to new opportunities to further define and promote this particular avenue of competitive advantage for individual firms.

Continued evidence of knowledge transfer has been seen in the marketplace leadership of area companies. For example, publicly-held Herman Miller, Inc. (Zeeland, MI), Steelcase, Inc. (Grand Rapids, MI) and privately-held Haworth, Inc. (Allegan) were among the first companies in the country to embrace cradle-to-cradle protocol (McDonough & Braungart, 2002) in product design, and Leadership in Energy and Environmental Design certification for facilities. Privately-held Cascade Engineering (Kentwood, MI) has been a leader in workforce innovation and is currently moving into design, production and licensing of residential-scale wind turbines.

The habit of collaboration for sustainable development deepened beyond the firm level when the West Michigan Sustainable Business Forum (WMSBF) was launched in 1994 to share best practices among like-minded companies. The group was an outgrowth of concerned business professionals associated with the West Michigan Environmental Action Council (WMEAC), at that time one of the area's leading environmental organizations. WMEAC served as WMSBF's fiduciary until the end of 2008 when WMSBF moved toward establishing itself with its own non-profit status. At the time of its initiation, the WMSBF was path-breaking with no other similar organization existing in the country. Today, the WMSBF has approximately 80 dues-paying member companies who meet monthly to exchange best practices. WMSBF recently broadened its focus by launching a 'regional coalescence' project to help leverage the strength of the business community with that of a broader collection of interested individuals charting a regional path toward sustainability.

Through the Business and Institutional Furniture Manufacturer's Association (BIFMA), competitors began a collaborative learning process in 2006 to develop nationally-recognized, industry-based sustainability standards specific to the office and institutional furniture industry (BIFMA, 2008). While individual companies, especially larger office furniture manufacturers such as Steelcase, Herman Miller, and Haworth have already been leveraging supply chains toward sustainable business practices, the new standards, now nearing final acceptance, will provide greater assurances of adherence, as well as greater penetration of triple bottom line thinking in the marketplace as a competitive strategy. More than 68 stakeholder groups (although not all regional) came together in this process.

Although market demand originally led a number of businesses to move toward adoption of sustainable business practices, supply chain leverage has led to an infusion of practices among smaller suppliers both locally, regionally and nationally. The Michigan Department of Environmental Quality's (MDEQ) administration of the U.S. Environmental Protection Agency's Green Supplier Network has helped to enhance learning in area supply chains. In fact, one of the MDEQ representatives working on behalf of the Green Supplier Network played a significant role in the BIFMA industry standard.

The region's emerging approach toward sustainable business appears to be undergirded by successful community collaborations initiated by business, as well as significant 'civic venture capital'[5] from area philanthropists. Contributions from the region's philanthropic sector have supported educational and civic-minded endeavors to move the region toward sustainability. Additionally, public commitments to sustainability made by the current mayor of the City of Grand Rapids, who has elevated regional awareness of sustainability, and has promoted the region's efforts nationally.

Although the heart of the sustainable business activity has been centered within the office furniture industry and its value chain, more recent efforts focus toward development of an alternative energy sector as a manufacturing diversification strategy for the region (The Right Place and WMSA, 2007). A $15 million, 3-year grant awarded by the U.S. Department of Labor, Employment and Training Education in 2005 allowed for the execution of a strategy called WIRED (Workforce Innovation in Regional Economic Development), boosting attention to sustainable innovation as one of the region's strategic imperatives. While much of the funding is dedicated to innovation in general, the hiring of two professionals, one devoted to sustainability and the other to advanced production technology by The Right Place, one of the region's economic development entities, has raised the profile of this perspective among other regional business leaders. Additionally, Design West Michigan, a WIRED innovation, endeavors to bring design thinking to area businesses of all types, deepening innovation competency in the region.

Grant monies have also supported a Sustainable Manufacturers Users Group, sponsored by The Right Place, the regional economic development agency, facilitated by a private sustainability consultancy, Sustainable Research Group, and co-hosted with Aquinas College, a private liberal arts college located in Grand Rapids. Approximately 21 individuals from seven of the area's larger companies have participated in a year-long monthly exchange of knowledge regarding sustainable business practices. A second cohort of seven companies continues in this program.

Higher education plays a critical role in sustainable innovation (Huggins, Jones, & Upton, 2008). Significant extension of a learning network gained momentum through philanthropic support in 2005, in the form of a 5-year, $1 million grant from the Steelcase Foundation, with additional 5-year support from the Wege Foundation. These gifts supported the contribution of other donors which helped launch the country's first undergraduate Bachelor's of Science degree in 2003 at Aquinas College. A number of recent graduates have found employment locally as sustainability professionals; interns have been placed at some of the region's largest employers as well as with small and medium enterprises; and representatives from local leaders in sustainable business serve on an external advisory committee. Meanwhile, students and recent graduates of the program are developing their own informal support network across the country to compare

---

[5]The concept of 'civic venture capital' was brought to the author's attention by Mr. Milt Rohwer, President of the Grand Rapids-based Frey Foundation.

sustainable innovation strategies. Other area academic institutions have also developed sustainability-related curriculum, such as vocational training in Leadership in Energy and Environmental Design (LEED) construction, wind energy and more general sustainability education.

A notable and recent development is also the establishment of the Michigan State University Bioeconomy Institute, located in Holland, MI in a facility donated by Pfizer when that company left the area. It opened in March 2009 to develop and commercialize plant-based products, as well as offer a biotech business incubator, in partnership with the area's economic development agency, Lakeshore Advantage.

Philanthropic support has also funded the Center for Sustainability at Aquinas College which serves as a link between the Aquinas academic program and the regional business community. In addition, the Center serves a regional connector role, offering a community sustainability resource and portal for regional sustainability initiatives through its website. Non-credit educational programs, consulting services, and the involvement of Center staff and affiliated professional in various community initiatives are also serving to transfer knowledge at regional scales. The Center has partnered with a variety of businesses to help facilitate learning about sustainable innovation at the firm scale.

The support of the Wege Foundation, as well as the personal philanthropy of Mr. Peter Wege, a member of one of Steelcase's founding families, has also been instrumental in sustainability-related education and partnerships more generally. For example, the Foundation is currently funding development of a public high school dedicated to 'Economicology^{TM}', creating a next generation of business leaders who will merge environmental stewardship with economic prosperity.

Political leadership has also been evident. Public support for the principles of sustainability has led various individuals to convene as groups focusing at local levels in Muskegon, Holland, Grand Haven and other areas in order to contribute to the overall awareness, understanding and acceptance of sustainability as a viable approach for community development.

8.4.4 Outcomes

Over the past two decades, West Michigan has seen continuing momentum in the growth of a diverse network of firms and individuals engaging innovation to support the development and deployment of sustainable business practices. Drivers of this activity have been market pressures, deeply rooted faith-based values, a strong outdoor heritage and a continuing commitment to reinvent the West Michigan community in response to changing conditions.

Learning derived from activity occurring at the level of the firm has been transferred through few formal organizations including the West Michigan Sustainable Business Forum, and Design West Michigan. More significant, however, is a learning network which is distributed in nature. In this sense, there is no formal, central operations or one leadership. Nodes of the network, including individuals, business firms and other organizations, have emerged but operate in a variety of spaces.

These links between various projects launched by formal organizations, supplant a centralized model of leadership, and offer instead what is called here as 'roving' leadership. The sustainable business network appears to be connected principally by key individuals, who link through their service to project-specific or focused, limited-scope efforts rather than constituting a formal organization or formally-constituted collaboration.

Norms supporting collaboration exist throughout the region. Over time, learning associated with sustainable innovation within individuals firms has led to a regional economic development focus rooted in sustainability, supported by political and community leaders. Institutions within the network appear to offer strong incentives for collaborative action and network structures are fostering sustainable innovation as a strategy for regional economic development at a variety of levels.

8.4.5 Evaluation

As noted above, the evaluative criteria in this case focus on the openness of the existing networks to both intra-regional and inter-regional learning in order to support sustainable regional development. Specifically, the significance of the networks' level of connectivity, their level of centrality and reachability (Janssen et al., 2006) can be brought to bear in understanding these activities in West Michigan region.

In terms of network connectivity, there appears to be an increasing density of the connections over time. Much of this connectivity appears to be strengthened through the continuing diffusion of innovation and sustainability within the region as valued approaches to business strategy and development. Inter-regional connections are fostered by professionals in key public and private companies, who continue to move forward as leaders within their industries, as well as at national and international levels. This exposure furthers not only the reputation of their individual firms but also highlights the penetration of sustainable business practices and innovation in the region.

The global affiliations and exposure of regional executives outside of the region expand the reachability of the network, countering regional lock-in through these connections. Key firms in the office furniture industry have a global presence and their reputation as leaders in sustainable business is well-known on an international basis. The firms' connections with other sustainable business thought leaders across the globe has encouraged an openness to new and emerging ideas.

In looking at the level of the centrality of the network, as noted above, distribution of information may be reduced if the centrality is high. This is perhaps where the West Michigan network may encounter some brittleness. There are several key individuals who serve as critical links within many of the nodes of the West Michigan sustainability landscape. If those individuals and others who are interacting with them conform to norms of open exchange, transparency, and accountability, then the experiences and ideas which can be exchanged will continue unfettered. However, if a particular individual impedes the free flow of ideas throughout the network by not passing along information once it reaches that individual from another node, then there is a risk of failing to capture the full benefits of a network's diversity (Janssen et al., 2006), p. 7). Additionally, attempts to disconnect critical links, even if only through neglect, may undermine the region's potential for sustainable business innovation.

While network assets are strong in the region's business community, assets for other sectors need to be identified and encouraged in order to continue efforts to move toward sustainable regional development. Other initiatives do continue to emerge, such as the recent designation of the area as one of the United Nation's Regional Centers of Expertise in Sustainable Development Education, and the WMSBF's 'regional coalescence' noted above. Critical to these regional efforts will be intentional actions to counter regional lock-out by assuring an awareness of the regional landscape and the networks which exist within it. Social networking tools may be key to this, helping to assure reachability.

8.5 Conclusions/Areas for Further Research

This research has illuminated the presence of a significant West Michigan 'network of networks' dedicated to sustainable business. It is not a conventional model of leadership, where a vision has been developed, a strategy designed and a team assembled to implement the strategy. This collection of overlapping networks has emerged to respond to the increasing pace of change and the disruptive circumstances which present themselves as a result of that change. It is a network open to learning and the transfer of knowledge relating to sustainable business.

A deeper understanding of this new model of networks as well as the interactions within the networks will help guide businesses, policy-makers and citizens as they attempt to rethink the human-environment relationship and reconsider possibilities for shaping sustainable regional development. Sustainable regional development should be an expected response to disruptive circumstances. West Michigan's early experiences toward regional development based on its experiences as industry innovators poises the region for success as firms scale up their knowledge through sustainable business network linkages, and expand those linkages beyond the sector of commerce.

Additional research may help to more fully understand how to better harness the private sector's growing commitment to sustainable business as a way to foster sustainable regional development. In particular, a better understanding of the roles played by individuals in facilitating or impeding sustainable innovation would be helpful. In this regard, policymakers need to be mindful of supporting old mindsets which tend to seek centralization and efficiency as hallmarks of progress.

Cross-scale linkages, those individuals and mechanisms which allow interactive learning efforts to 'scale up' and 'scale down', need to be better documented and understood by researchers and practitioners in order to foster policy support for these linkages. These linkages are critical in mediating the various facilitating and impeding contextual factors that will arise in moving toward sustainable regional development.

Mechanisms of accountability are also critical. Policymakers who put a stake in the ground and claim to be a sustainable region are essentially undertaking a simple act, and such a claim may be widely communicated. But the true measure of success will require evaluating regional outcomes against the principles of sustainability, and adhering to an on-going commitment to fostering innovation and openness to new ways of learning.

Innovating toward sustainable regional development offers a new opportunity to better align our social and ecological systems. Such a timely and significant endeavor is worth a comprehensive community effort. There appears to be significant value in the ability to collaborate, learn from each other, and embrace an openness to new configurations that could facilitate travels along a pathway which human society is only now beginning to see.

Acknowledgments The author gratefully acknowledges the research assistance of Aquinas undergraduate students Kalee Mockridge, Colin Knue and Chris Jacob, as well as the assistance of Calvin College colleague Gail Heffner, Ph.D., who helped with the facilitation of the cognitive mapping of West Michigan's sustainability landscape by participants at the October 2007 'summit' of the Community Sustainability Partnership. It is also noted that the author is actively involved in a variety of sustainability and sustainable business efforts in the West Michigan community and draws upon information and insights derived from those experiences in this research.

References

Amin, A. (2004). Regions unbound: Toward a new politics of place. *Geografiska Annaler, 86,* 33–44. Retrieved June 1, 2008, from http://eprints.dur.ac.uk/archive/00000073/01/Amin_regions.pdf.

Andersson, M., & Karlsson, C. (2004). *Regional innovation systems in small and medium-sized regions: A critical review and assessment* (Centre of Excellence for Studies in Science and Innovation (CESIS) Electronic Working Paper Series, Paper No. 10). Retrieved June 2, 2008, from www.infra.ktch.se/cesis/research/workpap.htm.

Anderies, J. M., Janssen, M. A., & Ostrom, E. (2004). A framework to analyze the robustness of social-ecological systems from an institutional perspective. *Ecology and Society, 9,* 18. Retrieved May 26, 2005, from http://www.ecologyandsociety.org/ vol9/iss1/art18/main.html.

Berkes, F., & Folke, C. (Eds.). (1998). *Linking social and ecological systems: Management practices and social mechanisms for building resilience.* Cambridge: Cambridge University Press.

Borup, M. (2005). *Approaches of eco-innovation: Uncertainty assessment and the integration of green technology foresight and life-cycle assessment as a policy tool.* Paper presented at the 11th Annual International Sustainable Development Research Conference, Helsinki, 7–8 June, 2005. Retrieved May 27, 2008, from http://www.risoe.dk/rispubl/SYS/syspdf/sys_7_2005.pdF.

Broekel, T., & Meder, A. (2008, June). *The bright side and dark side of cooperation for regional innovation performance* (Papers in Evolutionary Economic Geography #08.11). Retrieved June 10, 2008, from Utrecht University, Urban & Regional Research Centre Web site: http://econ.geo.uu.nl/peeg/peeg.html.

Business and Institutional Furniture Manufacturer's Association (BIFMA). (2008). BIFMA Sustainability Standard E3-2008-Draft. Retrieved March 20, 2008, from http://www.bifma.org/public/SusFurnStdArchive/Draft/2008-06-06_BIFMA_e3-2008.pdf.

Castells, M. (2000). *The rise of the network society.* Oxford: Wiley-Blackwell Publishers.

Chaminade, C., & Vang, J. (2006). *Globalization of knowledge and regional innovation policy: Supporting specialized hubs in developing countries* (Working Paper 2006/15). Retrieved

June 28, 2008, from Lund University, Center for Innovation, Research and Competence in the Learning Economy Web site: http://www.circle.lu.se/publications.

Christensen, C. M. (1997). *The innovator's dilemma: When new technologies cause great firms to fail*. Boston, MA: Harvard Business School Press.

Christensen, C. M., & Raynor, M. E. (2003/2006). *The innovator's solution: Creating and sustaining successful growth* (Collins Business Essentials Edition). New York: HarperCollins.

Cohen, M. W., & Levinthal, D. A. (1990). Absorptive capacity: A new perspective on learning and innovation. *Administrative Science Quarterly, 35,* 128–152.

Council on Competitiveness. (2008). Thrive: The skills imperative. *Report from the Compete 2.0 Series of the Council on Competitiveness*. Retrieved February 2, 2009, from http://www.compete.org/publications/detail/472/thrive/.

De la Mothe, J., & Paquet, G. (1998). Local and regional systems of innovation as learning socio-economies. In J. de la Mothe & G. Paquet (Eds.), *Local and regional systems of innovation* (pp. 1–16). Boston, MA: Kluwer Academic Publishers.

Elkington, J. (1997). *Cannibals with forks: The triple bottom line of 21st century business*. Oxford: Capstone Publishing.

Esty, D. C., & Winston, A. (2006). *Green to gold: How smart companies use environmental strategy to innovate, create value, and build competitive advantage*. London: Yale University Press.

Faber, H., Jorna, R. J., & Van Engelen, J. (2005). The sustainability of 'sustainability': A study into the conceptual foundations of the notion of 'sustainability'. *Journal of Environmental Assessment Policy and Management, 7,* 1–33.

Frondel, M., Horbach, J., & Rennings, R. (2007). End-of-pipe or cleaner production? An empirical comparison of environmental decisions across OECD countries. *Business Strategy and the Environment, 16,* 571–584.

Gibbs, D. (2000). Ecological modernisation, regional economic development and regional development agencies. *Geoforum, 31,* 9–19.

Gibson, C., Andersson, K., Ostrom, E., & Shivakumar, S. (2005). *The samaritan's dilemma: The political economy of development aid*. Oxford: Oxford University Press.

Grand Rapids Community Foundation. (2001). The Social Capital Community Benchmark Survey. Retrieved June 16, 2008, from http://www.cfsv.org/communitysurvey /mi3d.html.

Green, K., & Randles, K. (2006). At the interface of innovation studies and industrial ecology. In K. Green & S. Randles (Eds.), *Industrial ecology and spaces of innovation* (pp. 3–27). Northampton, MA: Edward Elgar.

Gunderson, L. H., & Holling, C. S. (Eds.). (2002). *Panarchy: Understanding transformation in human and natural systems*. Washington, DC: Island Press.

Haas, P. M. (2004). Addressing the global governance deficit. *Global Environmental Politics, 4,* 1–16.

Hawken, P. (2007). *Blessed unrest: How the largest movement came into being and why no one saw it coming*. New York: Viking/Penguin Group.

Hawken, P., Lovins, A. B., & Lovins, L. H. (1999). *Natural capitalism: Creating the next industrial revolution*. Boston, MA: Back Bay Books.

Hall, J., & Vredenburg, H. (2003). The challenges of innovating for sustainable development. *MIT Sloan Management Review, 45,* 61–68.

Huggins, R., Jones, M., & Upton, S. (2008). Universities as drivers of knowledge-based regional development: A triple helix analysis of Wales. *International Journal of Innovation and Regional Development, 1,* 24–47.

Inkpen, A., & Tsang, E. (2005). Social capital, networks and knowledge transfer. *Academy of Management Review, 30,* 146–165.

Janssen, M. A., Bodin, Ö., Anderies, J. M., Elmqvist, T., Ernston, H., McAllister, R. R. J., et al. (2006). A network perspective on the resilience of social-ecological systems. *Ecology and Society, 11*(1), Article 15. Retrieved May 22, 2008, from http://www.ecologyandsociety.org/vol11/iss1/art15/.

Johnson, M. W., Christensen, C. M., & Kagermann, H. (2008, December). Reinventing your business model. *Harvard Business Review, 86*, 51–59.

Keane, J., & Allisson, J. (1999). The intersection of the learning region and local and regional economic development: Analysing the role of higher education. *Regional Studies, 33*, 896–902.

Kelley, T., & Littman, J. (2001). *The art of innovation: Lessons in creativity from IDEO, America's leading design firm.* New York: Doubleday Publishing.

Könnölä, T., & Unruh, G. C. (2007). Really changing the course: The limitations of environmental management systems for innovation. *Business Strategy and the Environment, 16*, 525–537.

Lundvall, B. A. (1992). *National systems of innovation: Toward a theory of innovation and interactive learning.* London: Pinter Publishers.

MacKinnon, D., Cumbers, A., & Chapman, K. (2002). Learning, innovation and regional development: A critical appraisal of recent debates. *Progress in Human Geography, 26*, 293–311.

McDonough, W., & Braungart, M. (2002). *Cradle to cradle: Remaking the way we make things.* New York: North Point Press.

Meeus, M. T. H., Oerlemans, L. A. G., & van Dijck, J. J. J. (1999). *Regional systems of innovation from within: An empirical specification of the relation between technological dynamics and interaction between multiple actors in a Dutch region* (Working Paper 99.1). Retrieved June 13, 2008 from Eindhoven Centre for Innovation Studies, The Netherlands Web site: http://fp.tm.tue.nl/ecis/Working%20Papers/eciswp4.pdf.

Miguélez, E., Moreno, R., & Artis, M. (2008). Does social capital reinforce technological inputs in the creation of knowledge? *Evidence from the Spanish regions* (Working Paper 13). Retrieved March 20, 2009, from University of Barcelona, Research Institute of Applied Economics Web site: http://ideas.repec.org/p/ira/wpaper/200813.html.

Morgan, K. (1997). The learning region: Institutions, innovation and regional renewal. *Regional Studies, 31*, 491–503.

Murphy, J. T. (2006). Building trust in economic space. *Progress in Human Geography, 30*, 427–450.

Nicholson, N. (1993). The state of the art. In V. Ostrom, D. Feeny, & H. Picht (Eds.), R*ethinking institutional analysis and development: Issues, alternatives, and choices* (pp. 2–39). San Francisco, CA: Institute for Contemporary Studies.

North, D. C. (1990). *Institutions, institutional change, and economic performance.* New York: Cambridge University Press.

OECD. (2005). Organization for Economic Co-operation and Development: The measurement of scientific and technological activities: Proposed guidelines for collecting and interpreting technological innovation data ('Oslo Manual'). Retrieved June 16, 2008, from http://www.oecd.org/dataoecd/35/61/2367580.pdf.

Ostrom, E. (1995). Designing complexity to govern complexity. In S. Hanna & M. Munasinghe (Eds.), *Property rights and the environment: Social and ecological issues* (pp. 33–45). Washington, DC: The International Bank for Reconstruction and Development/The World Bank (and Beijer International Institute of Ecological Economics, Royal Swedish Academy of Sciences-Stockholm).

Ostrom, E. (2005). *Understanding institutional diversity.* Princeton, NJ: Princeton University Press.

Ostrom, E., Gardner, R., & Walker, J. (1994). *Rules, games and common-pool resources.* Ann Arbor: The University of Michigan Press.

Ostrom, E., & Janssen, M. A. (2002). *Beliefs, multi-level governance, and development.* Paper prepared for delivery at the 2002 Annual Meeting of the American Political Science Association, Boston, MA, USA, August 29–September 1, 2002.

Porter, M. E. (1985). *Competitive advantage: Creating and sustaining superior performance.* New York: The Free Press/Macmillan.

Powell, W. W., & Grodal, S. (2005). Networks of innovators. In J. Fagerberg, D. C. Mowery, & R. R. Nelson (Eds.), *The Oxford handbook of innovation.* Oxford: Oxford University Press.

Prahalad, C. K., & Krishnan, M. S. (2008). *The new age of innovation: Driving co-created value through global networks.* New York: McGraw Hill.

Rainey, D. L. (2006). *Sustainable business development: Inventing the future through strategy, innovation and leadership.* Cambridge: Cambridge University Press.

Rossi, M., Charron, S., Wing, G., & Ewell, E. (2006). Design for the next generation: Incorporating cradle-to-cradle design into Herman Miller products. *Journal of Industrial Quality, 10,* 193–210.

Schumpeter, J. (1934). *Theory of economic development.* Cambridge, MA: Harvard University Press.

State of Michigan. (2009). Michigan's December jobless rate increases. Press release prepared by the Department of Energy, Labor & Economic Growth. Retrieved April 1, 2009, from www.michigan.gov/dleg/0,1607,7-154-10573_11472-207233-,00.html.

Steketee, D. M. (2006). *Making connections: Environmental NGOs and cross-scale linkages in Ecuador's tropical forests policy process.* PhD Dissertation, Indiana University.

The Right Place. (2008, April). West Michigan Growth Statistics. Retrieved from www.rightplace.org.

The Right Place. (2009, January). West Michigan Fact Sheet. Retrieved from www.rightplace.org/regionalstatistics/.

The Right Place & West Michigan Strategic Alliance (WMSA). (2007, December). Alternative and Renewable Energy Cluster Analysis: A Growth Opportunity for West Michigan. Grand Rapids, MI: The Right Place. Retrieved June 1, 2008, from www.wm-alliance.org/alt_energy_cluster_analysis_Final.pdf.

Van Bakel, J., Loorbach, D., Whiteman, G., & Rotmans, J. (2007, December). *Business strategies for transitions towards sustainable systems* (Erasmus Research Institute of Management (ERIM), Report Series No. ERS-2007-094-ORG). Retrieved January 6, 2009, from http://www.hdl.handle.net/1765/10887.

West, E. (2008, October). America's greenest city. *Fast Company* 129 (On-Line Version). Retrieved November 1, 2008, from http://www.fastcompany.com/magazine/129/new-urban-eco-nomics.html.

West Michigan Strategic Alliance. (2002). The common framework: West Michigan/a region in transition. Retrieved June 13, 2008, from http://www.wm-alliance.org/documents/publications/The_Common_Framework.pdf.

Williams, R., & Markusson, N. (2002). *Knowledge and environmental innovations.* Paper presented at Blueprints for an Integration of Science, Technology and Environmental Policy workshop, 23–24 January, 2002. Retrieved May 28, 2008, from http://www.supra.ed.ac.uk/Publications/Paper_29.pdf.

WBCSD. (2007). World Business Council for Sustainable Development Annual Review. Retrieved June 12, 2008, from http://www.wbcsd.org/DocRoot/LsS9sBAiFBctMe3sjn4x/annualreview2007.pdf.

Zwaniecki, A. (2009). Grand Rapids, MI: Building on conservative values. Retrieved February 15, 2009, from http://www.america.gov/st/econ-english/2009/February/20090204135624saikceinawz0.3741114.html.

Chapter 9
Regional Perspectives on Capacity Building for Ecodesign – Insights from Wales

Simon O'Rafferty and Frank O'Connor

Abstract To contribute to SRD, regional and national governments will be required to support businesses and social enterprises in improving the sustainability performance of their products and services. There have been a number of national and regional programmes supporting ecodesign in SMEs but it is well documented that the implementation of ecodesign still remains low. Much of the literature has focussed on the organisational and methodological barriers to ecodesign. This chapter will contribute to this literature by highlighting the regional dimensions of ecodesign with a particular focus on ecodesign interventions for Small to Medium Sized Enterprises (SMEs). This exploration will highlight three key factors (a) that ecodesign can contribute to SRD, (b) that systems failure presents a rationale for regional interventions to enable ecodesign and (c) there is the need for a new dialogue on the structure and content of interventions supporting ecodesign in SMEs. To support this discussion, four SME case studies originating from a recent regional ecodesign initiative in Wales will be presented as a means to explore strategies for future interventions. This ecodesign initiative was delivered by the Ecodesign Centre (EDC) and supported by the Welsh Assembly Government (WAG).

Keywords Ecodesign · Incremental innovation · Small to medium sized enterprises · Regional support structures · Systems theory

9.1 Introduction and Context

Innovation is widely seen as a key mechanism for rapid regional economic growth, improving resource efficiency and moving towards sustainable regional development (SRD). Many regions attempt to foster innovation through public or semi-public infrastructure and private sector oriented policy measures. More

S. O'Rafferty (✉)
Ecodesign Centre, Cardiff Business Technology Centre, Cardiff, 24 4AY, UK
e-mail: simon@edcw.org

recently, regional policies are taking a systems perspective through developing an infrastructure for linkages and co-operation between actors and agents. There is an extensive body of literature addressing the interface between innovation and sustainability. Much of this literature focusses on the role of radical and systems innovation, such as innovation in energy or mobility systems. While much of this debate is essential, the cumulative impact of incremental innovations on long term economic development and social change can be equal or greater than radical innovations (DG ENTR, 2002; Fagerberg, Mowery, & Nelson, 2006; Kemp, Andersen, & Butter, 2004). This perspective opens up important avenues of discussion on the role of design in its various forms as a complementary asset for innovation and regional development.

Until recently, design as a creative process and business strategy has been under-represented in the innovation literature. Design has also been under-represented in the sustainable development (SD) literature. The general focus in the literature on technological innovations fails to prevent a complete picture of the role of design insofar as many innovations are based on novel designs or concepts as opposed to technical novelty (Tether, 2005; Whyte, 2005). While the third revision of the Oslo manual has extensive treatment of innovation outside of or ancillary to the development or use of technology, it remains limited in scope (OECD, 2005). The understanding of design is reaching beyond traditional perspectives on the design of products, services and brands towards more strategic considerations. The discussions have evolved from primarily ecological concerns to integrated discussions on sustainable consumption and production (SCP), social innovation and economic development in the broadest sense.

To contribute to SRD, regional and national governments will be required to support businesses and social enterprises in improving the sustainability performance of their products and services. There have been a number of national and regional programmes supporting ecodesign in SMEs but it is well documented that the implementation of ecodesign still remains low. Much of the literature has focussed on the organisational and methodological barriers to ecodesign. This chapter will contribute to this literature by highlighting the regional dimensions of ecodesign with a particular focus on ecodesign interventions for Small to Medium Sized Enterprises (SMEs). This exploration will highlight three key factors (a) that ecodesign can contribute to SRD, (b) that systems failure presents a rationale for regional interventions to enable ecodesign and (c) there is the need for a new dialogue on the structure and content of interventions supporting ecodesign in SMEs. To support this discussion, four SME case studies originating from a recent regional ecodesign initiative in Wales will be presented as a means to explore strategies for future interventions. This ecodesign initiative was delivered by the Ecodesign Centre[1] (EDC) and supported by the Welsh Assembly Government (WAG).

---

[1]EDC is an applied research organisation that aims to build capacity and capabilities to enable effective ecodesign. EDC was recently designated a Centre of Expertise by the WAG and is recognised as the 'voice' of, and knowledge base for, ecodesign in Wales.

9.2 Ecodesign Practice

Design has always been an inclusive process involving many specialists, communication channels and often large organisational structures. It is an increasingly fragmented and geographically diffuse activity that crosses international time zones and cultural barriers. Linear, staged and endogenous models have dominated research on product and service development. These models, while providing useful frameworks are increasingly insufficient in portraying the complexity of product and service development in the context of global supply chains, distributed manufacturing, disruptive innovation and ecodesign.[2] Design often has an exogenous organizational structure with complex relationships, distributed communication channels and multiple stakeholders representing potentially higher risk.

Within these models there are a number of management frameworks and tools that are geared towards providing insights on the outcomes or analytical processes of designing in a more sustainable manner. These frameworks are often challenging for designers and design managers as they incorporate processes and technical requirement outside of traditional design expertise. These include full life cycle analysis, full life cycle costing, new material considerations and increased standardisation. There are a number of areas that often remain overlooked in the literature such as adaptations needed for business organisations to put this knowledge into practice and the key capacities and competencies required by designers to implement these frameworks and tools.

One primary characteristic of the theoretical basis of ecodesign research to date is that a physical artefact or product is the focus of the ecodesign activity (Olundh, 2006). This physical artefact is placed in the context of the full life cycle of manufacturing systems (Fig. 9.1). Through consideration of this context, ecodesign aims to reduce or eliminate impacts of products and services (e.g. energy, materials, distribution, packaging and end-of-life treatment). The centrality of the product is indicative of the earlier research in ecodesign. For example, Brezet and van Hemel (1997) suggest that 'ecodesign considers environmental aspects at all stages of the product development process, striving for products that make the lowest possible impact throughout the product life cycle'.

9.2.1 Ecodesign and Sustainable Regional Development

Mirata, Nilsson, and Kuisma (2005) have offered a number of guiding principles for SRD, including, increasing the diversity and flexibility of economic activities, increasing wealth creation for a larger number of people, the sustainable use of local and preferably renewable resources, increasing the share of value added retained

---

[2]It is important to note that when the authors use the term ecodesign they include all perspectives on the role of design in SD, e.g. sustainable design, social design and potentially transformation design.

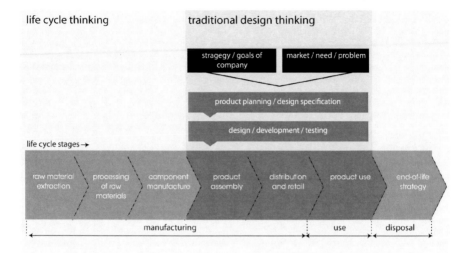

Fig. 9.1 Life cycle thinking

in the regions, increasing collaboration among regional activities while decreasing pollutant emissions and waste generation. If these guiding principles are distilled it suggests that SRD will address the move towards more sustainable patterns of production and consumption. While SCP is an inherently global concern the regional scale is perceived to be particularly important for the development of innovation systems that can facilitate the transition to SCP. This is generally accepted because of the spatial dimension of networks, knowledge spillovers and exchange, which are understood to lie at the heart of successful innovation (Cooke, Heidenreich, & Braczyk, 2004; Morgan, 2004).

Studies on the impact of design on firm performance and regional development are hindered by the lack of commonly agreed statistical measures. This can create difficulty when establishing causality and defining any correlation between design input and firm performance. Cereda, Crespi, Criscuolo, and Haskel (2005) assessed the impact that expenditure on design has on company performance using the 'Community Innovation Survey'. The study sought to cluster firms with the CIS according to their productivity and turnover to allow for an analysis of the correlation of design expenditure and performance. There appeared to be no univariate relationship between design expenditure and company performance but innovative firms investing in design tended to have higher growth. Haskel, Cereda, Crespi, and Criscuolo (2005) also considered design expenditure in a multivariate setting and found a positive correlation between the design expenditure and company productivity and growth. These findings are also supported by other research including that of the UK Design Council (Design Council, 2005).

Swann and Birke (2005) outlined the channels linking creativity and design to business performance in a basic framework. This builds on traditionally linear models through which R&D drives innovation, which in turn impacts company performance. While design can be a functional element of R&D, it also plays an important role in those sectors with little or no R&D activity. Tether (2005) argues that design activities link between the various categories of R&D while complementing all stages of the innovation process. Crucially, Tether has demonstrated, by means of the UK Innovation Survey of 2005, that design is an important functional asset for innovation. In the context of sector competition based on non-price characteristics, design becomes an increasingly essential complementary asset. The defining and ordering of these non-price characteristics becomes complex, as end-users will respond differently to the non-price characteristics. Within this particular discussion environmental performance can be defined as a non-price characteristic and the tangible consumer responses to environmental performance can be divergent.

A number of business benefits from applying ecodesign have been identified in the literature. These include improved brand, reduced production costs and improved product quality (Gertsakis, 2003; Gouvinhas & Costa, 2004; Núñez, 2006; Sherwin, 2004; van Hemel & Cramer, 2002; Wimmer, Züst, & Lee, 2004). Previous research by the authors identified a number of potential benefits for ecodesign as identified by Welsh SMEs (Fig. 9.2). In terms of the perceived business drivers these can be classified under three core areas; strategic (innovation, investment), internal (costs, competitiveness and social) and external (resource efficiency and communication). These strands of evidence would suggest that successful ecodesign could contribute to firm performance while offering positive externalities for SRD.

9.3 Theoretical Framework

9.3.1 Systems Failure

The discussion on rationales for intervention in economic systems has, until recently, been dominated by the market failure agenda. For policy makers, two conditions must exist for public sector intervention to be warranted. The first being that market mechanisms must fail to efficiently (or effectively) deliver on public policy objectives and, secondly, that intervention must lead to an improvement of the condition. From the neoclassical perspective, primary reasons for market failure include non-excludability of public goods, negative externalities, imperfect information and imperfect competition (GLA, 2006). It is understood that any intervention should provide 'additionality' and not replace a market function that would occur without the intervention having occurred. Therefore market failure is a necessary condition for intervention although it is not a sufficient condition in itself.

Recent discussions emerging from the evolutionary economics and innovation systems literature place a greater emphasis on systems failure as a rationale for intervention (Woolthuis, Lankhuizen, & Gilsing, 2005). In the context of this chapter the 'system' is defined as the innovation system. While there are many different contextual, theoretical, normative and temporal interpretations, the innovation

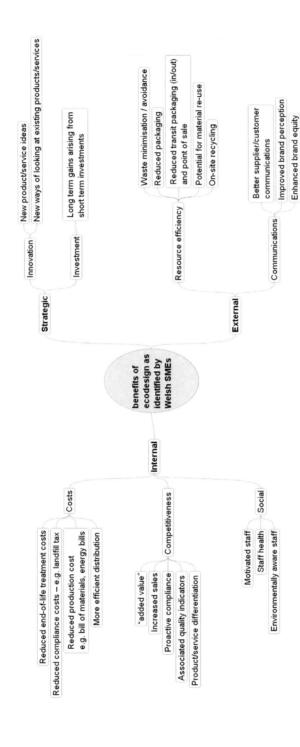

Fig. 9.2 Benefits of ecodesign as identified by Welsh SMEs

system broadly refers to the institutional, organisational, economic and socio-political factors that determine and diffuse innovations. An important condition of the innovation systems perspective is that it provides a framework through which the co-evolution of technologies, institutions and organisations can be analysed. It also explicitly infers that innovation is driven by the collaboration and interaction between different actors, institutions, knowledge flows and market conditions.

A number of authors have identified potential components and conditions that contribute to systems failure (Carlsson & Jacobsson, 1997; Edquist et al., 1998; Smith, 2000). These include failures in infrastructure provision and investment, lock-in and path dependencies, institutions, networks and capabilities. While the system failure rationale for intervention may be more intuitive for contemporary innovation scholars, the application of the concepts to the design of interventions has only very recently been addressed in the literature (Chaminade & Edquist, 2006; Woolthuis et al., 2005). It must be noted that in their discussion on public sector intervention in the context of innovation systems, Chaminade and Edquist (2006) avoided the use of the term 'failures' and instead used the term 'systemic problems'. They sought to avoid a possible misinterpretation of the systems perspective through seeking optimality or efficiencies in economic systems.

In the case of systems failure, the processes of intervention are similar in the case of market failure although the process is not focussed on recreating market conditions or optimum economic efficiency. Some of the key characteristics of systems failure interventions include increased collaboration and interactivity, a focus on learning and tacit knowledge, innovation capacity building, flexible and responsive policy frameworks and increased policy coherence. Scott-Kemmis, Holmen, and Balaguer (2005) identified a selection of key principles that may be adhered to when developing interventions based on system failure. The intervention should;

- Support international competitive analysis between sectoral innovation systems
- Provide mechanisms for increased collaboration, particularly in the area of public-private partnerships for the purpose of analysing opportunities, diagnosing problems, exploring options and strategic planning
- Facilitate learning and build company innovation capability
- Provide flexibility within the policy framework itself to ensure responsiveness to changes in the innovation landscape as interventions impact on the elements within the system

9.3.2 System Failure and Ecodesign in SMEs

In recent years there has been a decentralisation of European environmental policies and instruments. This has placed an increasing responsibility on regions to deliver SD (Burstrom & Korhonen, 2001). The principles of subsidiarity suggest that support and advice are most successful when delivered on the ground in a local context (Allman, Loynd, & Street, 2006). These factors have given rise to an increasing number of regional and local consulting programmes on sustainability issues for companies across Europe. There have been a number of programmes and interventions related to the implementation of ecodesign in SMEs. Some of these programmes were explicit and independent activities while some were part of larger programmes of sustainable business development. While these programmes took many forms (O'Connor & O'Rafferty, 2005; Tukker & Eder, 2000), the primary mechanisms of intervention included the provision of information services, demonstration projects, R&D financing, grants, establishing co-ordination bodies and 'brokering' services.

It is widely accepted that the implementation of ecodesign in SMEs still remains low. The literature has presented a broad range of barriers to ecodesign implementation, and more generally product development in SMEs (Belmane, Karaliunaite, Moora, Uselyte, & Viire, 2003; Jönbrink & Melin, 2008; Maxwell & van der Vorst, 2003; Millward & Lewis, 2005; Nauwelaers & Wintjes, 2003; O'Rafferty, O'Connor, & Cox, 2008; Rennings, Kemp, Bartolomeo, Hemmelskamp, & Hitchens, 2004; Tukker, Ellen, & Eder, 2000). Based on this empirical evidence of barriers to ecodesign implementation the authors present a framework of evidence towards systems failure as a rational for ecodesign intervention (Table 9.1). This is based upon a systems failure framework put forward by Woolthuis et al. (2005).

Table 9.1 Systems failure framework

Category	Failure
Infrastructure	Low representation of ecodesign indicators in government R&D programmes
	Low levels of investment in ecodesign related R&D
	Inadequate numbers of ecodesign support providers
	Low awareness by firms of emerging ecodesign related issues in key markets
	Lack of exposure to formal and informal ecodesign education and training
	Lack of alignment between ecodesign providers and industry
	Low utilisation of external knowledge providers
	Lack of support for intermediary organisations to build capacity in ecodesign
	Unclear market signals and demands
Institutions	Actors can not or will not act due to uncertainty and poor appropriability
	Competing policy rationales (e.g. environment and innovation)
	Government information asymmetries
	'Public-good' nature of investment
	Lack of policy supply and demand coherence leading to uncertainty and investment inefficiencies
	Regulators inflexible and too slow to change
	Regulators lack resources and expertise to address ecodesign issues
	Time lag between R&D intervention and commercialisation

Table 9.1 (continued)

Category	Failure
Interaction and networks	Little structured co-ordination of public-private partnerships or triple helix networks
	Lack of external support (training, advisory services, etc.) to develop ecodesign led innovations
	Organisational thinness in innovation and ecodesign support
	Lack of information on potential markets (niches)
	Limitations of the local markets (too small, low expenditure)
	Fragmented value chain structures
	Low levels of collaboration between technology commercialisers, international partner and R&D providers
SME capability	Fragmented product development process in SMEs
	Lack of managerial and operational resources
	Failure of managers to harness the strategic considerations
	Lack of viable technology options or alternatives
	Lack of awareness of viable technology options
	Lack of clear internal ecodesign or innovation strategies
Culture	Lack of top management commitment
	Lack of awareness, training, and motivation of employees
	Sustainability (environmental and social) viewed as periphery of core business
	Poor perception of ecodesign by investors
	Risk averse attitudes and resistance to engaging in new business opportunities through ecodesign
	Low levels of trust in intermediary and business support organisations
	Focus on short-term investments

This presented evidence of systems failure is coupled with an understanding that many of the previous public interventions did not facilitate second order additionality or sustainable changes in SME ecodesign practice. Previous research by the authors identified potential reasons for this including limited project scope, restrictive budget cycles, fragmented support mechanisms and lack of consideration of broader institutional contexts (O'Connor & O'Rafferty, 2005). This research also highlighted that the perceived low level of additionality may be related to how the intervention was monitored and evaluated. Difficulties in evaluation of these interventions include attribution of intervention to additionality and spill-over effects, time-lag between commercialisation and intervention, the measurement of qualitative effects such as improved absorptive capacity, competencies of SMEs and improved networks (Boekholt & Laroose, 2002).

This research by O'Connor and O'Rafferty (2005) also suggested that these interventions were based upon linear, neoclassical interpretations of innovation. This approach assumed that knowledge is generic, codified, immediately accessible and directly productive and that there is no difference between capabilities, knowledge and information (Hauknes & Nordgren, 2000). This allowed interventions to occur without consideration of the wider institutional context. This linear model of innovation is broadly contested and has given way to the recognition that innovation is an interactive, dynamic and non-linear process. It is therefore important to establish a framework of analysis that can incorporate the richness and interactivity of innovation.

9.4 Methodology

The exploratory empirical data consists of four SME case studies conducted within an ecodesign initiative supported by the WAG. The initiative was delivered over 18 months (11 March, 2007–10 September, 2008) by EDC. This ecodesign initiative incorporated ecodesign demonstration projects with four Welsh SMEs from priority sectors, ecodesign related research, evidence gathering to inform business support provision and WAG policy and where possible, testing new approaches to ecodesign intervention that aim to better meet the needs of Welsh SMEs. The companies were selected through a competitive multi-stage assessment process. A 'Design for Growth' study identified 205 SMEs, from a sample of 2056 (39% of all respondents), across Wales with a self-specifying design capacity that had indicated strategic and operational priorities related to the environmental and social impacts of their business (LEED, 2007). Another important consideration was the potential scale of environmental improvement and the scope and opportunity for second order additionality through knowledge spill-over and transferable models of implementation.

The data for these case studies has been collected through a number of methods including the 'Design for Growth' study results, project proposal forms, meeting diaries, company feedback forms/mechanisms, semi-structured interviews and questionnaires. This triangulation of the data allowed for a more rich foundation to the case development and analysis. A total of nine semi-structured interviews were carried out with representatives from different functional areas (e.g. director, production manager, senior design engineer). The representatives formed part of the SME projects teams that liaised with EDC. The semi-structured interviews consisted of direct and indirect questions that related to company activities prior to and during the intervention, strategic and operational characteristics, internal communications and interim perspectives on post-intervention additionality. In the analysis of the results the interviewees are treated anonymously in order to preserve confidentiality.

9.4.1 Initiative Process Model

The initiative process model highlights some key stages in the development and delivery of the ecodesign initiative and demonstration projects (Fig. 9.3). This model indicates how the intermediate and final outcomes are expected to address the objectives of EDC and the wider policy remit of the WAG. This model is an adaptation and simplification of a logic model. Logic models are an instrument used in

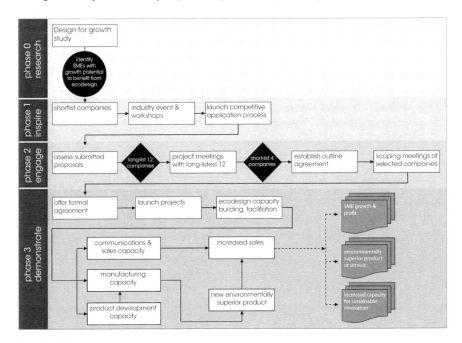

Fig. 9.3 Initiative process model

theory-based evaluations (Chen, 2005; Lipsey, 1993). The programme theory reflects the intended theory of action and the model is an articulation of this assumed theory (Patton, 1997). From this perspective, these models can serve to visualise the theory of action by displaying a logical sequence of steps and cause-and-effect mechanisms from programme implementation to expected outcomes.

Phase Zero: Research: It was important that companies most likely to benefit from ecodesign were identified at the outset. EDC worked in partnership with Cardiff Business School in carrying out a unique 'Design for Growth' study. This survey utilised a robust framework, the "potential growth index", to identify successful firms likely to benefit from ecodesign intervention.

Phase One: Inspire: Key growth companies and other priority stakeholders from the public and private sector were invited to an awareness-raising event and workshops organised by EDC. The objective of the event was to deliver a clear message to industry regarding the commercial benefits of ecodesign, promote and present the concept of the demonstration project, encourage participants to apply for the package and to act as a gateway to the selection process. The facilitated workshops allowed for a deeper exploration of the motivations and perceived competency gaps for ecodesign amongst design-led Welsh SMEs.

Phase Two: Engage: The submitted project proposal forms were evaluated against a set of criteria to establish the potential feasibility of proposals and to generate a shortlist of eligible companies. Ten companies were long-listed for the second phase of assessment. Individual company meetings were held to discuss proposals and conduct further detailed in-house assessment.

Through this process four companies were provisionally short-listed for the demonstration projects. A pre-selection meeting was then held with each, enabling the refinement of the project proposals and the drawing up of individual project plans and agreements.

Phase Three: Demonstrate: This phase consisted of a bespoke package of support. Each SME had ongoing access to EDC's in-house team, which provided specialist guidance, facilitation, research, monitoring and promotion. Interactive learning and collaboration was facilitated through 'Commercial Support Partnership' sessions. These sessions provided a platform for the SMEs to share experiences and transfer knowledge while receiving relevant technical and commercial know-how from industry experts. The companies also received up to £20,000 financial support that was assigned to specific ecodesign activities.

Each of the four companies followed the same process path as outlined above. Divergence occurred within the process during Phase three. The high degree of heterogeneity in the projects required a flexible and evolutionary intervention model. This allowed for a proactive but responsive framework of intervention that could accommodate different capacities and rates of change within a 12-month timeframe. Figure 9.4 outlines a broader perspective on the implementation model.

This intervention model is delineated along three dimensions; strategy, inputs and process. The strategy dimension is primarily concerned with the governance and management of the intervention, inputs relates to specific inputs to the intervention via EDC and process relates to specific activities that occurred throughout the demonstration projects. This Fig. 9.4 represents a systems view of the intervention model when viewed retrospectively.

9.5 Preliminary Results

This section presents the four company case studies and the analytical framework[3] for evaluating ecodesign interventions. Table 9.2 is an outline model of the analytical framework and it depicts the dimensions of inquiry for each case study. It has been developed and adapted from a framework developed by Morgan (2005) through his work on capacity building in the development context. The variables that make up the analytical framework are drawn from direct observation of the ecodesign intervention and in the capacity building, innovation and ecodesign literature.

The analytical framework contributes to this discussion through consideration of several aspects of ecodesign intervention. Firstly, it highlights some key aspects related to competencies for ecodesign while considering wider institutional contexts

---

[3]This analytical framework forms part of a more comprehensive analytical framework that will be reported upon through a PhD by Simon O'Rafferty.

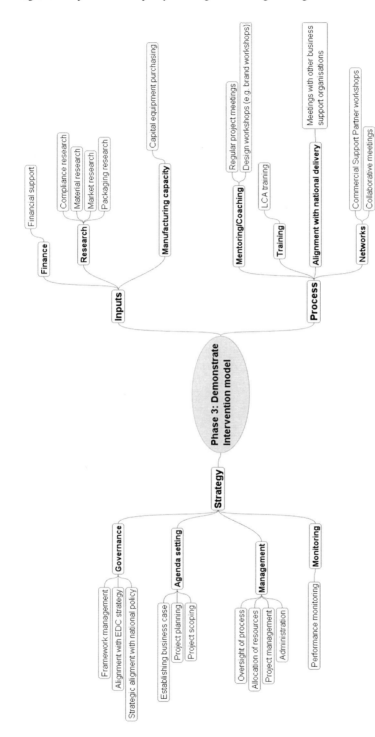

Fig. 9.4 Intervention model

Table 9.2 Analytical framework

Independent variables	
External context	Political context
	Institutional context
	Stakeholders (Suppliers, Customers, Clients, Retailers, Employees, Shareholders, Subsidiaries, Community)
External intervention	Interface of intervention
	Influence of external agents
	Capacity of intervention organisation
Dependant variables	
Capacity	Vision & strategy
	Intelligence & insight
	Culture, values & beliefs
	Competencies
	Organisation & process
Endogenous change and adaptation	Organisational change
	Process change
	Procedure change
	Product change
Performance	Effectiveness
	Change
	Measurement
Interaction	External collaboration
	Engagement

and levels of interaction. Secondly, the framework indicates several outcomes that may emerge from intervention such as product and process differentiation. One of the underlying objectives of the analytical framework is to aid more effective evaluation and development of ecodesign interventions in the context of systems failure.

9.5.1 Presentation of the Cases

The four companies that were selected were broadly representative of priority sectors in Wales. Sectors represented included electrical and electronic equipment (EEE), food and drink, low-carbon technology and manufacturing. The selected companies also represented a stratified sample of SME sizes; from a micro-SME up to an upper-range SME with over 200 employees.[4] Boxes 9.1, 9.2, 9.3 and 9.4 present a brief description of each company, its relation with the environment and initial outcomes.

---

[4]Micro-SME: Company A = 0–10 employees, Mid-range SME, Company B = 10–50 employees, Company C = 50–100, Upper-range SME, Company D = 100–250 employees. This classification is not official and was developed for this chapter.

Box 9.1 Case Study Company A

Company A (Micro-SME)

This company is a high-quality producer of a range of wholesome fresh soups, pasta sauces, ready-to-eat salads and sandwich fillings for both trade and retail sectors. Every product is manufactured to organic standards and is approved by the Soil Association.

Company A and the Environment

The food and drink sector has come under renewed scrutiny in recent years due to food scares, health conscious consumers and the environmental impact of food and packaging waste. The company has undertaken a number of initiatives to improve the environmental performance of their business. As the company is a micro-SME they faced a number of significant resource barriers to following through on elements of their strategy.

Initial Outcomes

Company A have developed their eco-friendly packaging and greener logistics alongside a brand and communications overhaul. The company have ambitious plans for growth. The packaging solution developed through the course of the project is not only more environmentally friendly but is more cost effective, allows for better portion control and increased shelf life. This has the potential to significantly reduce food waste for the consumer.

Box 9.2 Case Study Company B

Company B (Mid-Range SME)

Company B is a highly innovative, forward thinking company that design and manufacture specialist laser diode modules. The company are an Original Equipment Manufacturer producing products for a wide range of applications including machine vision, alignment, medical, measurement, and scientific.

Company B and the Environment

The environment was not a strategic or operational priority for Company B, their customers or the laser diode module sector until the last few years. The company had addressed onsite treatment and separation of waste. There were a number of reasons for this including scale, product characteristics and market conditions. The company has been consistently quality-led and this is evident in their ISO 9000 certification.

Initial Outcomes

Company B has developed their new 'green flagship' product (low power consumption laser). The results from the Life Cycle Analysis indicated that the new design has reduced the carbon footprint by up to 51%. This is because the new design uses 50% less energy than other products in their range. Because the company operates in a highly competitive business-to-business market, maintaining customer loyalty is crucial to their ongoing success. They recognise that creating a communications strategy around their new green flagship product can provide a useful means of differentiation from the competition and ultimately lead to increased sales.

Box 9.3 Case Study Company C

Company C (Mid-Range SME)

Company C is a global leader in the production of solar energy. The company manufactures dye-sensitised solar cells through a combination of innovative material science and nanotechnology. Their initial market will be mobile consumer products such as mobile phone chargers, smart textiles (incorporating the technology into fabrics), emergency applications, MP3 players and handheld game consoles.

Company C and the Environment

The environment and Sustainable Development are at the core of the company's business strategy. This is reflected in the products they make and the manner in which they produce them. The company aims to be the first manufacturing facility to develop renewable products solely through the use of renewable energy. The company has been exploring the potential for onsite energy generation to enable carbon neutral manufacturing. Thin-film solar technology is far more resource efficient than conventional photovoltaic systems, and Company C are using silicon and cadmium-free processes to reduce the cost of manufacturing.

Initial Outcomes

The company has developed its first PV-integrated product, a mobile phone charger. While these products can be developed in house, an important element of the company's approach is to develop strategic partners for future product development.

The strategic focus of Company C remains on emerging markets (India, Brazil, China), low income sectors of developed and transition economies and what is known as the base of the pyramid model (the 4 billion individuals that live primarily in developing countries and whose annual per capita incomes fall below $1,500 (PPP).). Serving these markets profitably requires companies to rethink many aspects of their strategy and operations. There are significant challenges for a company at the front-end of Research and Development (R&D) to align speed to market with the development of new markets and the stabilisation of the technology.

Box 9.4 Case Study Company D

Company D (Upper-Range SME)

Company D is a UK market leader in the research, development, manufacture and service of seating for the commercial environment. They design and manufacture for the UK and European business to business markets. The company's key strength is responding to changes in the workplace and enabling businesses, and their working environment, to be flexible. The company has undertaken a number of award winning environmental initiatives.

Company D and the Environment

Office furniture production tends to require material and components from a range of suppliers. This poses challenges when determining the full life cycle impacts of the product. The greatest environmental impacts are generally in the production phase resource consumption (raw materials, coatings, solvents and water), production phase emissions (emissions to air, effluent, solid waste), production phase energy consumption, transportation and end-of-life (EOL) treatment.

Initial Outcomes

Through their initial project proposal the company indicated an interest in undertaking a Cradle to Cradle (C2C) design project. The C2C protocol is a life cycle orientated approach to design that considers the development of materials and products that are safe and suitable for recycling through technical or biological systems. One of the key considerations throughout the C2C process was the sustainability of the process without a functioning infrastructure for the EOL management of their products. EDC and Company D initiated a programme of research and capacity building around the potential for an EOL system to support any C2C activity in the company. These activities are still ongoing. The environmentally superior product features include product light weighting through a mono-material backing unit, improved assembly and disassembly times and improved overall resource efficiency.

9.5.2 Initial Analysis

This subsection will provide some initial analysis of the case studies based on the proposed analytical framework.

The four companies displayed different strategic perspectives on ecodesign. Three of the companies were already addressing sustainability issues, either as a codified strategy or embedded within the company culture. In the case of the micro-SME the director was passionate about sustainability and was driven to champion the issues through the products and business communications. One of the mid-range SMEs was strategically driven towards sustainability but a number of factors were seen to potentially limit the scope of this strategy. For example, this company was still within a rapid stage of R&D and strategic partner development. The internal perception amongst the project team was that ecodesign required a process of reflection and exploration and this was seen to conflict with the quick pace of change and the need for speed-to-market.

In the case of the upper-range SME, sustainability was already a key element of the core business strategy. In this case it was an evolving agenda because a number of related initiatives had taken place within the company although none of these previous activities were paradigm changing or strategically 'sticky'. The project team displayed a great deal of openness and flexibility to change and learning. Although not explored in detail, the historical context of the company may have had an important role in shaping these vales of flexibility and openness. This particular company originated from a management buy-out of a subsidiary to a larger multi-national task furniture company that has similar strategic characteristics and values.

It is useful to retrospectively observe the role of tangible assets such as technology, existing finances and staff on the ecodesign projects. The companies displayed a great deal of variance in available tangible assets that they could allocate to the projects. Three of the companies were able to allocate different members of staff to the projects over the course of the 12 months. While allowing for the introduction of new perspectives and skills it also increased the resilience of the projects within the companies. One of the companies faced significant staff restructuring throughout the course of the project and this had implications for the rate of change and delivery.

9.5.2.1 Capacity

Capacity is a multi-dimensional concept incorporating competencies, capabilities and values. While generally viewed in technological and financial terms it is also concerned with managerial, organisational and relationship-based competencies. Companies will often draw on existing capacities when implementing ecodesign as it is often regarded as a process of continuous improvement or incremental innovation. On the basis of the theoretical and empirical insights, it would appear that an optimum level or equilibrium of capacity can be difficult to define or specify. The notion of optimality may also run contrary to the evolutionary perspective.

Because optimality is difficult to define it would suggest that there is a dynamic interplay across the dimensions of capacity. This interplay can be observed through the absorption of new capacities through the exploratory and exploitation phases of the intervention. The insights from evolutionary perspectives on innovation would suggest that the adoption and retention of capacities that can be utilised in different contexts is important. In two of the cases there was exploitation of learning and dormant competencies from previous interventions that had not been applied commercially in a new ecodesign project. This highlights the importance of understanding the cumulative impact of capacity building activities.

9.5.2.2 Endogenous Change and Adaptation

The intervention model required the formation of project management teams drawn from different functional areas of the businesses. These project teams acted autonomously following senior management sign-off on the project proposals. While the intervention model was largely responsible for defining this organisational structure and interface, a selection of ad-hoc endogenous changes over time were observed. For example, two of the companies integrated the ecodesign project into the regular internal design reports and company R&D meetings. This allowed for the formalisation of discussion on environmental and sustainability issues across various levels of the business. This also ensured that the ecodesign activities remained on the company agenda throughout the course of the project.

A key area of interest was how the initial focus on incremental product innovations could be a vehicle for broader innovations across the business. In terms of input additionality, the ecodesign projects induced innovation investment in two of the businesses. In one case this was explicitly linked to more sustainable innovations related to individual producer responsibility and company wide resource efficiency. In another case, the ecodesign project required a minor investment in new production equipment requiring new competencies and procedures. These endogenous changes were not mandated at the start of the intervention and were facilitated gradually over a number of months.

In general, the endogenous changes were facilitated through consensus building and collective learning supported by EDC. In the case of the two mid-range SMEs the collective learning focussed on specific skills development and training. The upper-range SME presented a very different form of endogenous change and adaptation. The design team held a strong autonomous position within the business and displayed leadership in developing the agenda within the business before engaging in consensus building. From the perspective of the intervener this process was more participatory and the learning reciprocal.

9.5.2.3 Performance

Performance relates to the way in which the intervention enabled the SME to apply new and existing competencies. It is therefore about execution and implementation. It must be stressed that the link between intervention and performance is difficult to measure as there are a number of intangibles and trade-offs to consider. The time lag between intervention and commercialisation can also disrupt robust analysis. As suggested by Morgan (2005), performance should be viewed as an emergent pattern that comes about through the interactions of many contextual elements both internal and external. This would suggest that the performance was inherently dependant on the other variables in the analytical framework.

The authors would suggest that the heterogeneity of the project teams gave rise to divergence in the applied intervention model and therefore performance varied between the SMEs. In all of the cases the intervention had both direct and indirect effects on the performance of the project team. These effects on performance related mostly to improving strategic direction, assisting internal consensus building, specific technical competencies and external communications. The depth of consensus on the performance of the intervention across all levels of the business was not clear but it was suggested that good performance would provide a mandate for future ecodesign activity.

A common issue in public sector programmes is the deterioration and decay of performance during the post-intervention period. There were a number of mechanisms used to reduce the potential of performance deterioration in the future such as benchmarking and organisational learning. The effectiveness of these mechanisms can only be measured through longitudinal analysis.

9.5.2.4 Interaction

The intervention model provided for interaction and partnership development on a number of levels, including internally to the business, business to business, business to intervener and business to other intermediary organisations. These partnerships were integrated and developed incrementally to allow for the formation of trust and consensus based relationships. It was evident that the level of value gained from the interactions was not consistent across the companies. For the micro-SME the identification of value and learning was more challenging. The company director suggested this was due to scale, sector and context. One of the SMEs did not fully engage with the partnership process.

The intervention provided a platform for interactive learning through a series of 'Commercial Support Partnership' sessions. The companies represented diverse sectors with individual needs. The non-competitive nature of the session initially allowed for supportive interactions. It was suggested by a number of representatives that the value of the sessions was the building of confidence on the scope and intent of the ecodesign projects. Over time, the potential for longer-term commercial collaborations and partnerships was discussed openly but the implementation of these was outside of the timeframe of this study. It must also be noted that one of the companies suggested the economic downturn was the major barrier to investment and development of new partnerships.

It is widely understood that interactive and inter-company learning are a fundamental attribute of market economies and that many companies draw their knowledge from external sources. The interaction between these businesses required structured co-ordination as they would not have seen any value in interacting with each other before. This would suggest that future processes of intervention should focus on a longer term strategy of building knowledge infrastructure aimed at innovative capacities on the regional or national level.

9.6 Discussion and Conclusions

This chapter sought to contribute to the literature by addressing the regional dimension of ecodesign in SMEs and as such was inspired by three issues. The first issue was the potential role of ecodesign in supporting SRD. The second was the role of systems failure in providing a rationale for intervention. The third was a need for a new dialogue on the structure and content of interventions supporting ecodesign in SMEs in the context of systems failure. To contribute to this new dialogue the chapter reflected on the complexity of developing regional interventions supporting ecodesign in SMEs. The analytical framework was informed by the literature but the relationships among variables were oriented towards ecodesign intervention. The framework captured the formal, informal, internal and external influences on capacity building for ecodesign including interactions between SMEs and the public sector.

The evidence provided through the case studies suggest that SMEs require a flexible and evolving intervention model that can compensate for a lack of structured coordination of ecodesign activities. The intervention requires guiding structures, both formal and informal, while not being overly prescriptive. There is a potential risk of an overly rigid and linear intervention model giving rise to the formalisation paradox. This formalisation paradox describes the situation where actors seek to formalise processes and activities that are generally intuitive and open, leading to static and rigid development. This would run counter to the often intuitive and non-linear process of product design and development in SMEs.

Another important insight is that the complexity underpinning ecodesign intervention in SMEs is born out of the interdependency between the internal and external contexts of ecodesign practice. The case studies highlighted a high degree of interdependency but also a degree of ambiguity in the nature of the intervention and the roles of the intervener. This ambiguity is driven by a combination of historical factors such as previous interventions experienced by the participants but also the interactive nature of the intervention in question. In many instances the learning is reciprocal between SME and intervener and between each company involved in the programme. The case studies also highlight the potential complexity of meso-level interventions that seek to address the content and direction of micro-level design practice. This complexity is primarily driven by the idiosyncrasies of SME organisation and behaviour and the multi-dimensional nature of ecodesign.

The authors propose using a capacity building perspective to inform interventions for ecodesign as a response to the system failure framework outlined in this chapter. The concept of capacity building does provide a useful vocabulary by which the dynamic context of ecodesign and ecodesign interventions can be explored. Capacity can be defined as the combination of competencies, capabilities, knowledge, networking opportunities and motivations that are required to create value. Creating value in this context is applying ecodesign.

Despite the differing local contexts, organisational characteristics and performance measures, the case studies suggest a number of common capacity building themes that should be built into future intervention strategies. These include using and developing local knowledge networks and partnerships, facilitating longer-term trust-based relationships between businesses and intermediary organisations, improving supply and demand side policy coherence and providing inspirational platforms that allow for interactive learning. Design, and by extension innovation, is fundamentally a collaborative process therefore these intervention strategies reinforce the implicit theme of collaboration running through this chapter. The need to break old social and organisational silos while creating new collaborative contexts for design and innovation is increasingly important in the context of sustainable development. The recent convergence of global economic crises and renewed sustainability concerns is in part due to the networked nature of economic systems. While this may be the case, it is networks and the active participation in networks through collaboration that will drive design and innovation for sustainability in the future.

The issues under discussion in this chapter form a component of a broader research programme on capacity building for ecodesign in SMEs.[5] While the evaluation and analysis is still at an early stage it is anticipated that these initial discussions will contribute to the debate on the potential role for ecodesign in SRD and the importance of collaboration for innovation and sustainability.

References

Allman, L., Loynd, A., & Street, P. (2006). *Bringing innovation to existing approaches to influence the behaviour of small businesses: Final report by the National Centre for Business and Sustainability for the Department for Environment, Food and Rural Affairs*. Manchester: The National Centre for Business and Sustainability.

Belmane, I., Karaliunaite, H., Moora, R., Uselyte, & Viire, V. (2003). *Feasibility Study for Eco-Design in the Baltic States Industry, 2003–2004*, Copenhagen. http://www.norden.org/pub/ebook/2003-559.pdf.

Boekholt, P., & Laroose, J. (2002). *Innovation policy and sustainable development: Can innovation incentives make a difference?*. Brussels: IWT-Observatory.

Brezet, H., & van Hemel, C. (1997). *Ecodesign: A promising approach to sustainable production and consumption*. Paris: UNEP.

Burstrom, F., & Korhonen, J. (2001). Municipalities and industrial ecology: Reconsidering municipal environmental management. *Sustainable Development, 9* (1), 36–46.

Carlsson, B., & Jacobsson, S. (1997). Diversity creation and technological systems. In C. Edquist (Ed.), *Systems of innovation – Technologies, institutions and organization* (pp. 130–156). London: Routledge.

Cereda, M., Crespi, G., Criscuolo, C., & Haskel, J. (2005). *Design and company performance*. London: DTI.

Chaminade, C., & Edquist, C. (2006). Rationales for public policy intervention in the innovation process: A systems of innovation approach. http://www.proact2006.fi/chapter_images/303_Ref_B207_Chaminade_and_Edquist.pdf.

Chen, H. (2005). *Practical program evaluation: Assessing and improving planning, implementation, and effectiveness*. Thousand Oaks, CA: Sage Publications.

Cooke, P. N., Heidenreich, M., & Braczyk, H. (2004). *Regional innovation systems: The role of governances in a globalized world*. London: Routledge.

Design Council. (2005). *Design index: The impact of design on stock market performance*. London: Design Council.

DG ENTR. (2002). *Innovation tomorrow innovation policy and the regulatory framework: Making innovation an integral part of the broader structural agenda*. Brussels: DG ENTR.

Edquist, C., Hommen, L., Johnson, B., Lemola, T., Malerba, F., Reiss, T., et al. (1998). *The ISE policy statement – The innovation policy implications of the 'Innovations Systems and European Integration'*. Research project funded by the TSER programme (DG XII). Sweden: Linkoping University.

Fagerberg, J., Mowery, D. C., & Nelson, R. R. (2006). *The Oxford handbook of innovation*. Oxford: Oxford University Press.

---

[5]EDC are conducting a longitudinal evaluation of the industry demonstration projects.

Gertsakis, J. (2003). *Facilitating EcoDesign and product stewardship in Australia: A review of EcoDesign and product stewardship interventions emanating from the Centre for Design at RMIT University (1990–1999)*. Melbourne: Centre for Design at RMIT University.

GLA. (2006). *The rationale for public sector intervention in the economy*. London: Greater London Authority.

Gouvinhas, R., & Costa, C. (2004). *The utilisation of ecodesign practices within Brazilian SME companies*. Global Conference on Sustainable Product Development and Life Cycle Engineering, Berlin, Germany.

Haskel, J., Cereda, M., Crespi, G., & Criscuolo, C. (2005). *Creativity and design study for DTI using the community innovation survey*. London: DTI.

Hauknes, J., & Nordgren, L. (2000). Economic rationales of government involvement in innovation and the supply of innovation-related services. Accessed November 6, 2007. http://econpapers.repec.org/paper/stpstepre/1999r08.htm.

Jönbrink, A., & Melin, H. E. (2008). *How central authorities can support ecodesign: Company perspectives*. Copenhagen: Norden.

Kemp, R., Andersen, M. M., & Butter, M. (2004). *Background report about strategies for ecoinnovation*. The Netherlands: VROM.

LEED. (2007). *Growth Study: Design in Wales, Final Report to Ecodesign Centre Wales*. Cardiff: Cardiff Business School. Prepared by Sally O'Connor (unpublished).

Lipsey, M. W. (1993). Theory as method: Small theories of treatments. *New Directions for Program Evaluation, 57*, 5–38.

Maxwell, D., & van der Vorst, R. (2003). Developing sustainable products and services. *Journal of Cleaner Production, 11* (8), 883–895.

Millward, H., & Lewis, A. (2005). Barriers to successful new product development within small manufacturing companies. *Journal of Small Business and Enterprise Development, 12* (3), 379–394.

Mirata, M., Nilsson, H., & Kuisma, J. (2005). Production systems aligned with distributed economies: Examples from energy and biomass sectors. *Journal of Cleaner Production, 13* (10–11), 981–991.

Morgan, K. (2004). The exaggerated death of geography: Learning, proximity and territorial innovation systems. *Journal of Economic Geography, 4* (1), 3–21.

Morgan, P. (2005). *Study on capacity, change and performance*. Maastricht: ECDPM.

Nauwelaers, C., & Wintjes, R. (2003). Towards a new paradigm for innovation policy. In B. Asheim, A. Isaksen, C. C. E. M. G. Nauwelaers, & F. Totdling (Eds.), *Regional innovation policy for small-medium enterprises* (pp. 193–220). Northampton: Edward Elgar.

Núñez, Y. (2006). *The Spanish eco-design standard. Implementation in SMEs*. Proceedings of CIRP International Conference on Life Cycle Engineering. Leuven, Belgium.

O'Connor, F., & O'Rafferty, S. (2005). *Developing a Welsh ecodesign initiative stage 1*. Interim Report submitted to the Welsh Assembly Government, Cardiff. (unpublished, contact authors for details), Cardiff, UK.

O'Rafferty, S., O'Connor, F., & Cox, I. (2008). *Supporting sustainable regional innovation and ecodesign in small to medium enterprises: A discussion on the issue with insights from Wales*. Proceedings: Sustainable Consumption and Production: Framework for action, 10–11 March, 2008, Brussels, Belgium.

OECD. (2005). *Oslo manual: Guidelines for collecting and interpreting innovation data* (3rd ed.). Brussels: OECD Publishing.

Olundh, G. (2006). *Modernising ecodesign: Ecodesign for innovative solutions*. Stockholm: KTH Industrial Engineering and Management.

Patton, M. Q. (1997). *Utilization-focused evaluation: The new century text*. Thousand Oaks, CA: Sage Publications.

Rennings, K., Kemp, R., Bartolomeo, M., Hemmelskamp, J., & Hitchens, D. (2004). *Blueprints for an integration of science, technology and environmental policy (BLUEPRINT)*, Mannheim: Centre for European Economic Research (ZEW).

Scott-Kemmis, D., Holmen, M., & Balaguer, A. (2005). *No simple solutions: How sectoral innovation systems can be transformed*. Canberra: Australian National University.

Sherwin, C. (2004). Design and sustainability. *The Journal of Sustainable Product Design, 4* (1), 21–31.

Smith, K. (2000). Innovation as a systemic phenomenon: Rethinking the role of policy. *Enterprise and Innovation Management Studies, 1* (1), 73–102.

Swann, P., & Birke, D. (2005). *How do creativity and design enhance business performance?* London: DTI.

Tether, B. (2005). *Think piece on the role of design in business performance*. London: DTI.

Tukker, A., & Eder, P. (2000). *Eco-design: Strategies for dissemination to SMEs. Pt. 2, Specific studies*. Brussels: Joint Research Centre, European Commission.

Tukker, A., Ellen, G. J., & Eder, P. (2000). *Eco-design: Strategies for dissemination to SMEs. Pt. 1, Overall Analysis and conclusions*. Brussels: Joint Research Centre, European Commission.

Van Hemel, C., & Cramer, J. (2002). Barriers and stimuli for ecodesign in SMEs. *Journal of Cleaner Production, 10* (5), 439–453.

Whyte, J. (2005). *Management of creativity and design within the firm*. London: DTI.

Wimmer, W., Züst, R., & Lee, K. (2004). *ECODESIGN Implementation: A systematic guidance on integrating environmental considerations into product development*. Austria: Springer.

Woolthuis, R., Lankhuizen, M., & Gilsing, V. (2005). A system failure framework for innovation policy design. *Technovation, 25* (6), 609–619.

Chapter 10
Fostering Responsible Tourism Business Practices Through Collaborative Capacity-Building

Bruce Simmons, Robyn Bushell, and Jennifer Scott

Abstract This chapter reviews two collaborative research and development projects in Australia, both focused on improved sustainability outcomes by small businesses. Each was funded by the NSW State Government, and built on established partnerships between university researchers and a local government body (Manly Council, New South Wales), and the researchers and a tourism industry association (Caravan & Camping Industry Association NSW). A theoretical model of engagement and state government policy on sustainable development underpins the approach used to analyze the case studies. Both the local government area and the caravan and camping industry have benefited from a number of financial incentives. Industry champions have been highly influential, while an extension model of capacity building and training has also contributed to the successful strategies. The common barriers to adoption of environmental management systems by industry operators who chose not to participate have been concerns about time, expertise, cost and bureaucracy. These case studies demonstrate the importance of collaborative partnerships; of context-specific strategies and approaches; and of adaptive management models to include the evaluation of both success and failure of process.

Keywords Barriers to adoption · Local capacity-building · Public–private partnerships · Sustainability planning · Management tourism business

10.1 Introduction

In the United Nations Decade of Education for Sustainable Development, 2005–2014 (UNEP, 2005) the fuzzy feel-good of the rhetoric of sustainability is gradually metamorphosing into a more defined shape, the edges hardening, the way

B. Simmons (✉)
School of Natural Sciences, University of Western Sydney, Locked Bag 1797, Penrith South DC, NSW 1797, Australia
e-mail: b.simmons@uws.edu.au

J. Sarkis (eds.), *Facilitating Sustainable Innovation through Collaboration*, DOI 10.1007/978-90-481-3159-4_10, © Springer Science+Business Media B.V. 2010

forward clearing. In Australia the need to hard-sell a complex set of abstract moral principles has been subsumed into a maelstrom of such momentum that many, especially in industry, have been galvanized into the need to take positive action. Energizing the community to act often requires a catalyst—and the growing evidence for global warming and the continuing conditions of severe drought appear to have penetrated the public consciousness as governments and businesses scramble to find ways to jump aboard the sustainability juggernaut.

Appropriate to the UNEP plan but developed prior to it was 'Learning for Sustainability 2002–2005,' the first 3-year environmental education plan developed by the government of the state of New South Wales, Australia. (The second plan, for 2007–2010, builds on the groundwork established by the first.) Under this plan, the overarching issues affecting the goal of sustainability in New South Wales were listed as follows.

- Sustainable lifestyles—sustainable consumption and production, social values and desirable futures, development and equity and global perspectives;
- Ecosystem health and bioregional awareness;
- Infrastructure and institutional arrangements;
- Local communities taking action—increasing community involvement in managing the environment, improving community access to better information, addressing the specific needs of community sectors including ethnic communities, Aboriginal and Torres Strait Islanders, regional and rural NSW, increasing Aboriginal involvement in managing National Parks and reserves. (NSW Council on Environmental Education, 2002, p. 10).

These issues are underpinned by a set of specific objectives which allow an integration of grounded, real world outcomes with the more esoteric arguments of sustainability. Such clarification has enabled educators to define the link between the feel-good arguments of sustainability to real-world, practical goals that can be modified to be relevant to any context, anywhere.

The Learning for Sustainability plan contains six principles to underpin sustainability learning, which have been used to guide the development and implementation of the strategy within the two cases discussed in this paper (NSW Council on Environmental Education, 2002, p. 9). The six principles are:

1. The development and delivery of environmental education in NSW (is) aimed at assisting the community to move toward sustainability;
2. Environmental education (is) integrated with other environmental management tools;
3. Environmental education acknowledges the complex connections between diverse aspects of environmental problems;
4. Environmental education promotes social change through the initiatives of individuals and organizations;
5. Environmental education is relevant to all aspects of our lives and is regarded as a lifelong learning process;

6. Continual improvement is at the basis of all planning, delivery and evaluation of environmental education.

Educating the community about sustainability has been difficult because of the range of preconceived ideas that surround this complex proposition. Educators in New South Wales, along with educators around the world, have struggled over the past few to years to engage both public and private sectors with the absolute necessity of taking account of sustainability issues in day-to-day activities.

In general, Australian business has been slow to adopt social and environmental responsibility, although it has been well recorded that 'big' corporations have responded more positively to community outrage by moving toward such responsibility. Surveys undertaken in Australia over the last 10 years have indicated a low rate of adoption of environmental management systems (a process for planning and monitoring performance for sustainability), in the order of 6–7% for small businesses, 20–23% for medium-sized businesses, and 17–35% for large business operations (Greene, 1999; Rynne, 2003, reported in Simmons et al., 2006).

The surveys, conducted in 1998 by the Co-operative Research Center for Waste Management and Pollution Control (Greene, 1999) and in 2002–2003 by the Australian Chamber of Commerce and Industry (Rynne, 2003, reported in Simmons et al., 2006) indicated a low initial rate of adoption of environmental management systems (EMS), with little improvement over the following 4 years. The barriers to adoption listed by the two surveys were similar, and can be summarized as lack of time, excessive cost, low priority, and lack of knowledge. The surveys revealed both a lack of understanding of the benefits of EMS, and a lack of realization of industry capability in the designing of EMS programs. This suggested that the level of awareness and education within specific industries was low, and that the EMS programs had been designed outside the industry sector and been imposed upon it. Improvements in both aspects fall to the individual industry sectors themselves to address.

Small to medium enterprises (SMEs) have been much slower to respond to community concerns because (among other reasons) they do not have the resources of big business to address the environmental and social values they may impact. Small businesses working on small profit margins consider that the survival of their business would be compromised if they had to carry the social and environmental costs they create. Under these circumstances, the government and hence the community have to carry this burden. The need therefore was to design and implement programs which concurrently address all the above concerns and to evaluate progress against the low adoption rates by SMEs around the world.

This article reports on two cases of sustainability learning programs that were trialed in SMEs. Both examples are associated with the tourism industry but operate in distinctly different management environments. Each case has its unique set of issues which on deeper examination demonstrate a remarkable similarity with common elements. One example is set in the private sector, the other in the public sector; both attempt to address a comparable set of conceptual and attitudinal problems.

The two cases encapsulate the principles of sustainability learning. Each program had, as a clearly defined fundamental goal, motivating its audience to make a shift from the 'business as usual' paradigm. Each applied a range of environmental and social tools to enable on-the-ground outcomes which generated a tangible difference in sustainability. No person, organization, or business exists in isolation from its community; hence links into the community were clarified and used to enable sustainability messages and methods to be extended both spatially and chronologically.

The cases exemplify the potential for effective social change processes. The synergies between the facilitators (the research partners) and the facilitated (the target groups) enabled them to merge and materialize as a true partnership in problem solving. An important, though perhaps seemingly obvious, finding in our research is that demonstrating the relevance of each component of the triple bottom line (TBL) is a matter of a thorough and credible understanding of the targeted community. TBL analysis is a way of relating social, environmental and economic outcomes—each of these elements has indicators of status that can be used to assess sustainable business performance (Global Reporting Initiative, 2002). TBL reporting by the target businesses provided them with greater understanding of their values and capabilities and capacity to achieve behavioral change.

10.2 Background

In the 1970 Boyer Lectures, prominent Australian economist H. C. Coombs commented that humanity functions on a base set of fragile patterns that have bred institutions to protect and promote those patterns. Institutions create systems to control activities—for example, the economic system. Originally designed to help producers be self-supporting, the economic system evolved to foster trade, export, monetary valuation, and speculation to guide production. Although conceptually a self-regulating system, the modern economic system requires frequent intervention to keep working correctly. Competition between the systems and their institutions fostered dominance and fashioned human behaviors to ensure that dominance continued (Coombs, 1970).

Coombs explains much about how and why business behaves in particular ways. Understanding the complexity, pressure points, and reality for a business in contemporary Australia allows the educator to better fashion 'hooks' and avoid 'sinkers' when attempting to engage business operators in the challenge of sustainability and TBL reporting. Coombs warned that systems eventually take on a life of their own. Thirty-five years later, social researchers such as Clive Hamilton and Richard Denniss concur with Coombs, confirming his theory that the economic system has acquired a disproportionate guiding influence over decision making, a sub-context that is readily evident in today's business institutions. Australia is deeply entrenched in and aligned to the 'growth at all costs' factor (Hamilton & Denniss, 2005).

The economic growth ethic is largely dismissive of, and blind to, the rapidly accruing cumulative damage that unchecked growth incurs (Smith & Scott, 2006).

There is, however, according to Richard Welford and his colleagues (Welford et al., 2006), a moral obligation for businesses to be involved in the process of sustainable development—Asia, for example, is at a crisis point of rapidly deteriorating environmental conditions and declining social equity, a crisis brought about by the schism between the path of sustainable development and the traditional Western economic growth model now so firmly inculcated into first world countries. Circumstances in Australia are quite different to those in Asia, however. Australia's relatively low population density has allowed a buffering of the negative impacts generated by the economic system on environmental and social values whereas Asia's high population densities allow no such cushioning effects.

Recent research into SMEs has found that they lack awareness of sustainability issues and lack the information necessary to connect their performance to their effect on sustainability (Viere, Herzig, Schaltegger, & Leung, 2006). SMEs lack the appropriate tools to evaluate their longer term performance and hence have short-term agendas focused squarely on income generation. The resources to pursue non-economic and indirect business pressures are scarce, often because of upper management's dismissive attitude to non-economic factors (Viere et al., 2006).

The scale and seriousness of Australia's social and environmental problems are no longer a contestable premise; the 'do nothing and allow business to continue as usual' approach is no longer an option (Lowe, 2005) and many larger corporations are now taking action. For example, leading bank Westpac has adopted the Global Reporting Initiative to provide feedback to stakeholders on the company's TBL performance (Business in the Community, 2006). The conundrum for SMEs which do appreciate the seriousness of the problem is that there are few practical models available that allow them to clarify the position of their business in relation to sustainability and corporate responsibility, or to take action in a logical and affordable manner. Many of the models offered to this audience are expensive to implement and impractical to maintain when resources are scarce (Simmons et al., 2006). SMEs claim they are time-poor, and lack both the resources and the capital to invest in such programs.

Sustainability requires a trans-disciplinary approach, which means that participating businesses need to form partnerships for the necessary skills to establish their performance relative to contemporary stakeholder expectations (Viere et al., 2006). Working collaboratively with stakeholders to develop and implement a program is the foundation of a successful sustainability initiative and a key message of this article. The program we discuss is based on a tiered system comprising engagement, process, and performance. It relies on developing both internal and external partnerships to appreciate the sustainability context for that business. Engagement is the foundation of the process. The quality of the subsequent outcomes depends heavily on the scope and adequacy of the partnership engagement. The examples involving the Caravan & Camping Industry Association and businesses in the Manly business precinct (a major coastal tourism destination in Sydney, the capital of New South Wales) demonstrate some of the difficulties encountered when working with SMEs to develop and implement sustainability plans and outcomes.

10.3 The Projects

10.3.1 Caravan & Camping Industry Association Gumnut Awards

The Caravan & Camping Industry Association of NSW (CCIA) represents some 410 holiday, tourist and residential park operators in New South Wales, Australia. Such parks are often located in areas highly vulnerable to human activities, for example beach dunes, river banks, wetlands and ridge tops. Many caravan park managers lack environmental knowledge and consequently were unaware of the impact their business was having on the natural assets of their local area. Equally they had little appreciation of the financial dependency of their business on those natural assets.

A partnership between the industry peak body, the CCIA, and the University of Western Sydney (UWS) was formed to develop an integrated environmental management and sustainability education system. The objective was to achieve environmentally and socially responsible business operations throughout the caravan and camping industry (Desailly et al., 2004). The pillars of the program were: ESD principles; self-development; capacity-building; and credible, independent accountability leading to a marketable business advantage. An awards system—known as the Gumnut Awards—was devised in conjunction with the program in an attempt to encourage its uptake. The three levels of award promote awareness, engagement, knowledge acquisition, systematic planning, monitoring, and performance assessment. The Gumnut Awards program is an accreditation scheme specifically designed for the camping and caravan park industry sector. It is currently available only to the members of the NSW association and has had a remarkably high adoption rate by park operators.

The Gumnut Awards program encourages CCIA members to reduce the environmental impacts of their parks, to explore mechanisms to increase benefits from their operation to local communities, and to transform their socially and environmentally responsible practices into economic advantage. Long-term changes in attitude, culture and practice are needed for acceptance and implementation of these concepts, and the CCIA has a realistic understanding of this and the associated investment required. The program is an industry association-driven process.

Each stage in the awards program is designed for applicants to identify, act on and monitor and review socio-environmental opportunities for their businesses. The Bronze Award is the first stage, based on a tailored training program designed to engage the operator into the program via a relative simple standardized self-assessment and desk audit with constructive feedback. This is followed by the Silver Award, which involves receiving assistance to put in place a business-specific planning mechanism that will lead to improvement in operator identified socio-environmental goals. Finally, the Gold Award rewards achievement via an ongoing iterative process that systematically addresses internal and external matters and monitors progress through an operator-designed environmental management plan.

At Bronze and Silver level workshops, operators share ideas and experiences for building capacity, and are given expert guidance and feedback to help develop simple, effective, flexible, site-specific and integrated systems. The United Nations Environmental Programme (UNEP, 1998) states that the main task of any environmental management program is to support organizations to improve performance utilizing tools, advice, close association and regular consultation. The Gumnut Awards recognize this high-level international priority, being designed to meet best practice standards relating to accreditation identified as part of the ongoing research into the awards and of research undertaken jointly by UNEP, the World Tourism Organization, and the International Ecotourism Society (Honey, 2002).

10.3.2 The Sea Change for Sustainable Tourism Program

Sea Change for Sustainable Tourism is a sustainability education program modified from the CCIA Gumnut Awards program and redeveloped specifically for tourism businesses operating out of the Manly business precinct (and complicated by the fact that Manly is also a residential suburb). Again, the program was designed to encourage the environmental and social performance of the tourism business sector, this time in one specific location involving a wide variety of business types, from accommodation providers, to restaurants, tour operators, attractions and a range of ancillary service providers. It was adopted in order to strike a balance between tourism, protection of the local environment, and the amenity of the residential community.

Sea Change for Sustainable Tourism was modeled on the highly successful CCIA Gumnut Awards and an existing environmental education program, Manly Council's Sea Change for Stormwater program. It was funded by the NSW Government's Environmental Trust and was an initiative of Manly Council in partnership with the University of Western Sydney (UWS) and the Manly Chamber of Commerce.

Sea Change for Sustainable Tourism adopted a five-star, tiered award approach involving self-assessment, sustainability planning and action. It also involved training workshops. Like the CCIA, Manly Council employed an education/extension officer to administer the program, liaise with stakeholders and support individual businesses. The ongoing program continues to be managed and funded by the Council (possibly with the assistance of State government grants). This program was a public agency-driven process which sought to develop a partnership between council and tourism business in the development and implementation of an accreditation program.

10.4 Evaluation of the Programs

Given that the national rates of adoption of environmental management systems by SMEs in Australia are in the order of 6–7% (Greene, 1999; Rynne, 2003, reported in Simmons et al., 2006), the rates of adoption of the two programs under evaluation appear extremely high. For the Gumnut program this amounted to 145 member businesses (or 35%) in the first year, and 205 member businesses (50%) of the association total of 410 businesses within 3 years. The Sea Change program engaged 27 (or 18%) of the 151 targeted local businesses by the end of the first year of operation.

There are a number of differences between the two programs (Table 10.1). The first of these is that the Gumnut program was driven by the industry itself through its association, while the Sea Change program was driven by a local government agency. For a local government to lead such a process creates an immediate difficulty in that resources (public funds) have to be accounted for and shown to be cost-effective to a wider constituency in an annual funding process. This is difficult in a program that can take a number of years to become established and successful. On the other hand, a business wearing the costs or effort can use internal benchmarks to show progress to members of the association and, provided the effort is not eroding business survival in the short term, can realize the benefits over longer than annual cycles.

Table 10.1 Comparisons between the Gumnut and sea change programs

Process	Gumnut awards environmental management program	Sea change for sustainable tourism
Constituents	Member caravan and camping parks in the same association	Local businesses in a tourism precinct
Driver	Industry association	Local government (council)
Program developed by	Industry with guidance from University of Western Sydney	Council (with guidance from University of Western Sydney) and industry consultation
Funding	Initial and educational funding by industry association Assessment funded by individual members	All funding by council and grants
Incentive development	Planned staged publicity of awards, and built into the ongoing action process	Opportunistic publicity, and built into the ongoing action process
Education process	Provided as awareness building, workshops, extension, and ongoing resource guidance	Provided as awareness building, workshops, extension, and ongoing resource guidance
Award levels	Three (bronze, silver, gold)	Five (one to five stars).
Number of categories	Ten (landscape, water & wastewater, solid waste, energy efficiency, air & noise pollution, biodiversity & conservation, economics, staff, community, safety & emergency response planning)	Ten (visual amenity, water, waste, energy, air/noise, biodiversity, economics, staff, community, risk management)
Auditing	Initial stage self-audit	Initial stage self-audit
Planning and action	Later stages of program as a result of the initial audit	Later stages of program as a result of the initial audit
Assessment	Independently assessed	Council assessed
Review	By industry association	By council

It is important, however, for local communities through their councils to set local environmental values and establish goals for sustainability. Intervention in the form of education programs is a common way to hasten behavior change toward sustainability; however, the changes may need to be supported in the long term by the target business.

10.5 Successes, Limitations and Barriers in Each Project

10.5.1 The CCIA Gumnut Awards

The Gumnut Awards program has attracted 50% of the 410 members of the association within 3 years of inception. Apart from this high uptake rate the project has created a substantial amount of data about parks and their management. The audit phase, for example, produced some useful data about the understanding park operators had of various aspects related to their business. It allowed the researchers to build a picture of the strong and weak areas of performance for the industry as a whole.

Within the audit phase participants were asked to rank their performance in each of the 10 categories (Table 10.4), firstly perceived against a set of criteria relevant to most parks, and secondly measured against a set of clearly defined benchmarks created by the industry itself. As Table 10.2 illustrates, there was a close relationship between the top three results and the CCIA benchmarks. Air and noise pollution was the strongest performing category, with staff and landscape management a close second and third. This indicates parks understood the issues the best and at least in the top three categories there was strong consistency between the two data sets.

In areas of weak performance, consistency between the two data sets is more varied in these results except for biodiversity. Table 10.3 shows that biodiversity conservation was a problematic concept for many park operators, and from comments given in the assessments it was clearly the outstanding issue when it came to understanding where the parks' responsibilities lay. The results have enabled the CCIA to developed targeted education strategies to strengthen performance in these weaker areas.

Changes over time in the attitude and capacity of park operators to effectively control impacts and develop extension strategies are being mapped within the Gumnut framework. Park operators must re-apply for award accreditation every

Table 10.2 Categories of strong performance

Ranking	Participants' perceived performance	CCIA measured benchmarks
1	Air & noise pollution	Air & noise pollution
2	Staff	Staff
3	Landscape	Landscape
4	Economics	Energy efficiency
5	Local community	Safety & emergency response planning

Table 10.3 Categories of weak performance

Ranking	Participants' perceived performance	CCIA measured benchmarks
6	Biodiversity conservation	Biodiversity conservation
7	Safety & emergency response planning	Solid waste
8	Solid waste	Local community
9	Energy efficiency	Water & wastewater
10	Water & wastewater	Economics

2 years and demonstrate they are continuing to improve on their performance. A framework has been established that allows ongoing monitoring of performance and encourages planning in the short, medium and long term.

The program is designed as an engagement model first and foremost, and secondly as a capacity-building tool for the industry. As an engagement model it has succeeded when compared to relative measures, but results have been slower to materialize in the capacity-building tool aspect. Under the continuing research program, park operators are invited to submit case studies of successful examples of projects they have undertaken as part of the Gumnut program.

A mail survey to all members was followed by a targeted telephone survey to encourage park operators to submit a case study. Twenty case studies (two in each of the 10 benchmark categories) was the target, with the aim of including more in areas defined as problematic for participating parks, and identifying reasons some parks had not yet engaged in the program. The mail survey produced returns of 14.5% for non-participating parks and 12% for parks already engaged. The results for those parks not yet engaged in the program suggested that operators were very much aware but had not participated because they:

- believed they were already engaged with sustainability issues;
- considered there was little value for them in engaging in the program;
- did not understand the Gumnut program concept.

A great deal of planning, time, energy, and resources has been devoted by the CCIA to engaging park operators in the program. These results suggest that more needs to be done to motivate the remaining 50% to engage with the program. While the majority of non-participating park operators commented that they intended to sign up for the program, this is yet to happen. Further thought needs to be given to better promoting the program to those park operators who claim they are interested but can't allocate the time to attend a workshop or be visited by the field extension officer.

While the Gumnut program has made substantial gains over the average results for programs with a similar aim, it still has to overcome some substantial prejudices associated with green strategies in business. Modifying the green image may be important to ongoing success.

10.5.2 Sea Change for Sustainable Tourism

Collaboration in the framework of public and private partnerships proved both valuable, and a barrier to successful outcomes, in the Sea Change for Sustainable Tourism program.

The first year's participation rate reflected only the last 6 months of the program, as the first 6 months were taken up with designing and creating the materials for program implementation. A number of workshops were conducted but attendance was constrained by the fact that they coincided with the busy season for the target businesses. The Sea Change for Sustainable Tourism (SCST) Steering Committee recommended that the program's Project Officer deliver the initial workshop material directly into the place of business to give an option to attending workshops.

At 12 months 27 (or 18%) of the 151 businesses targeted had engaged in the program. Despite the workshop constraints, this 18% uptake rate compared favorably with the 6–7% uptake rate in earlier Australia-wide programs with a similar focus on small businesses (Greene, 1999; Rynne, 2003, reported in Simmons et al., 2006). Two surveys were conducted to establish the success of the project and determine future issues and options for the program. A survey of 30 businesses (and stakeholders) participating in the program was conducted, with nine responses. The qualitative data was analyzed for primary themes and rationale; the results appear in Table 10.4.

In a second survey 32 non-participating businesses were contacted and 15 responded. Of these 10 indicated that they were aware of the Sea Change program. The non-participant survey also evaluated resistance to engagement. The reasons cited by businesses yet to enroll for delaying their entry into the program appeared consistent with earlier surveys of small to medium business (Greene, 1999; Rynne, 2003, reported in Simmons et al., 2006). These included:

- have not been approached
- selling the business
- no reason
- don't know enough about it
- not applicable to business
- no time

The qualitative data from the surveys suggest that the strength of the SCST program lies in its awareness and capacity-building components, but also suggest

Table 10.4 Sea change participant survey thematic data analysis results

Questions	Primary themes from responses	Rationale
Usefulness of the SCST program	Value lies in practical application and relevance to business	Businesses do understand the concept and value of triple bottom line performance in the tourism sector but need a practical way to adopt it
	Education opportunities welcomed.	Knowledge is valued
	Enthusiasm for ethical behaviors.	Businesses prefer to be seen as problem solvers, not problem makers
	Hidden rewards and opportunities available	Environmental and social responsibility has advantages
	Still early days; program must be given time to get known and valued	Continue program development; concepts and process should be periodically validated by stakeholders
Strengths of the SCST program	Awareness raising	Triple bottom line performance needs to be promoted as the contemporary indicator of a good business for customers and investors
	Time to consider importance of background issues	Issues relevant to the community have to be considered
	Project resources valued	Resources available to business accepted
	Simple, workable framework	Business people want commercially compatible strategies, not green motherhood statements
	Consistency with good business practice	Continuing development of resources needed; linkages with accepted business management principles
Weaknesses of the SCST program	Need for more grounded solutions to everyday problems	Good quality resources for ongoing extension capacity critical
	Incentives to participate need intensifying	Multiple levels of incentive, from easy to hard
	Continuing refinement and periodic reassessment by designers	Iterative design basis for continuing improvement by both businesses and designers
	Success not well enough promoted as an important marketing tool	Public awareness and perception highly valued; a critical mass of businesses in the program helps persuade undecided businesses to join
	Expanding from three to five levels of awards and the adoption of star symbols	More cumbersome and time-consuming; confusion with other star rating systems (e.g., accommodation)
	Collaborative partnership needed strengthening	A true partnership between the agency promoting the program and participating business is necessary for success
	Workshop timing was not always convenient	Business owners may have busy periods and cannot attend workshops
Future challenges for the SCST program	Council as extension providers	Entrenched traditional cultural norms and attitudes may limit potential

Table 10.4   (continued)

Questions	Primary themes from responses	Rationale
	Shorter versions to save time	Do it once, do it well and save money and time
	Learning from other businesses	Learning about and accepting a mix of solutions including expert, industry, local, and peer solutions
	Building collaborative partnerships	Difficult for regulators to convert to collaborators; need for a social learning basis for the program.
	Independent assessment and auditing	Needed for credibility of the program
	Commercial reality	Businesses pay for auditing to ensure independence of the program and life beyond funding; promotes commitment by business
Opportunities for the SCST program	Concept of a 'tourism' business	Remove barriers created by definitions
	Extension to other industries and programs	Arbitrary separation of business categories; advantages in developing critical mass
	Stakeholder groups not yet in the program	Cross-sector partnership roles and process need to be understood and accepted
	Celebrating success	Successful businesses need to be celebrated as leaders
	Valuing the program	Commercialization potential via public/private partnerships

that the program was lacking in meeting the specific needs of the target businesses. The results of these surveys will guide the continuing implementation of the program.

10.6 Reviewing Processes, Constructing Partnerships and Improving Theory and Practice

Based on the results of surveys of industry prior to 2004, the implementation of environmental management systems in small to medium business requires the following attributes to be successful (Simmons et al., 2006):

- principles of sustainability;
- development and operation of the program by the industry itself; and
- independent accountability

These points led to identifying a number of barriers that will need to be overcome to encourage for small business to adopt environmental management systems. These barriers are listed, along with the methods proposed to overcome them, in Table 10.5.

The adaptive method of the Gumnut and Sea Change programs proved valuable in capturing relevance and refining the program format to best suit the needs of their target audiences. Understanding the real-world dilemmas for a small or medium enterprise was critical to a successful outcome. In the caravan and camping industry, for example, pressures were brought to bear during the period of the study by a government authority concerned about the capacity of the parks to manage their impacts on the local environment. When viewed from a triple bottom line perspective, park operators who failed to engage with the requirements of good land stewardship and corporate responsibility ran the risk of falling foul not only of the local community, but also of government authorities and, most importantly, their customers.

Many leading park operators understood that good environmental stewardship and corporate responsibility were just as much the hallmark of a well-managed business as profit growth. The application of the Gumnut program provided a vehicle to engage with park managers and to incrementally construct a capacity within the industry to appreciate and effectively manage the park in a more environmentally and economically sustainable and socially acceptable manner.

The application of the model to the Manly Sea Change program proved somewhat more complex, as the process was condensed into a shorter timeframe. The results to date nevertheless demonstrated that it was imperative to integrate existing stakeholder knowledge about local conditions into a system of evaluation, action, and review for it to be effective. Local businesses understood the importance of the local environment to the area as a tourism destination. They also understood that those who benefit either directly or indirectly from the destination have a responsibility to contribute to the preservation and enhancement of local natural and cultural assets.

Table 10.5 Addressing barriers to adoption of environmental management systems in the Gumnut program

Barrier	Method proposed to overcome barrier
Lack of time	Planning by the industry itself and integration into existing activities
Excessive cost	Planning by each business within their budget constraints; recognition of actions already undertaken
Low priority	High level of industry and community publicity through association journals and public recognition of achievement; achievement used as a marketing tool
Lack of knowledge	Training workshops an essential part of the process
Lack of credibility	Independent assessment
Lack of commitment	Investment of time and costs in the program by participating businesses to promote ownership

Source: Based on Simmons et al. (2006).

It has taken an ongoing research process and partnership to direct and analyze components of the success of the implementation of environmental and social
responsibility in small and medium enterprises, and to identify the barriers to implementation. Addressing reported barriers in terms of relevance and capacity building
has led to some successes, in that greater numbers of businesses have engaged in
the sectors targeted. The initial cohort of businesses readily interested in participation appears relatively amenable to such programs. The next group was those
businesses that waited until they could see evidence of whether the program was
successful before jumping on board. The remainder were more difficult to shift.
These businesses have refuted or disregarded the mounting evidence surrounding
the benefits of participation, citing the reasons noted in Table 10.6 for their lack of
involvement/motivation.

For both programs the latter two reasons were common, and suggest that awareness of relevance and the advantages of participation in the program need to better
conveyed. While ownership of the programs lies with the industry and the local area,
not all businesses feel strongly connected to the networks that operate within these
settings. That connection is ultimately a telling aspect regarding the success of any
venture put forward by the collective group.

While Sea Change for Sustainable Tourism was developed from the Gumnut program, it required significant variations to account for the different context—that is:

- Manly as a destination with many different types of tourism businesses, not just
  one sector as in the case of the caravan and camping industry;
- the geographic proximity of each of the businesses in Manly compared to the
  distribution of members throughout New South Wales in the caravan and camping
  industry;
- the homogeneous socio-economic profile of Manly compared to the variation
  throughout the wider tourism regions of the State;
- the presence of a strong industry champion compared to working with local
  government employees.

These variations affected the ways in which the programs operated, the complexity of the partnerships, and their success. In addition, the CCIA had a significant
imperative for achieving successful results because of impending problems with
various State government agencies; the CCIA also has a highly motivated CEO who
fully appreciates both the need for EMS and the educational extension model of

Table 10.6 Reasons for resisting participation

Gumnut program	Sea change program
Already engaged in environmental activities	No time to devote to such a program
See little advantage for the business	Not applicable to the business
Don't understand the concept	Don't know enough about it

capacity-building to deal directly with issues rather than hoping the problem will go away.

The educational component of the programs has proven to be a vital aspect of the success of both, allowing participating businesses to understand what was required, why and how they could improve their viability, and at the same time operate more ethically. It is this capacity-building component that separates these programs from many others. The adaptive management nature of both programs has also enabled learning to be achieved by the program designers and the implementers, and for modifications to be made along the way. Through this the programs are more likely to continue to be successful. The partnership aspect of an independent third party, the university researchers, has provided a mechanism for the program leaders— the CCIA and Manly Council—to be audited and to be required to listen to their constituents concerning barriers to implementation and assistance required. This has been an important element of the success of the programs, and provides a further layer of capacity-building.

Knowing what is important to the stakeholders, identifying the hooks and sinkers in terms of engagement, was vital. Once engaged, the target communities benefited from an overarching framework that allowed them to map progress and reinvigorate the need for continual improvement. Nothing breeds success like success, and it was important in the early stages of the program to ensure that wins were rewarded both explicitly and implicitly. Public and peer recognition, and regular profiling of the success and integrity of participating businesses, has been reinforced implicitly through the economic bottom line, with direct financial savings from the reduction of waste and gains in market share. The explicit aspects, in terms of signage and awards, have significant marketing advantage and contribute to the financial benefits of becoming socially and environmentally responsible.

10.7 Conclusion

Government agencies in Australia have accepted the responsibility of encouraging sustainable development in the community, but attempts to promote sustainable behavior in small to medium business have been met with low adoption rates. Each business sector and business location can have specific processes and issues to which generic and imposed programs do not apply.

The programs reported in this article have demonstrated that success requires collaboration and cooperation with the target business in development and application. Lead agencies need to involve the target businesses in the development and delivery of innovative, change-embedded programs. Such programs require adaptation to the needs of the target business in their surrounding environment.

Out of this collaboration has emerged some useful indicators of the processes necessary to allow education and capacity building directed toward autonomous sustainable improvement by small to medium business.

References

Business in the Community. (2006). *Companies that count: Corporate responsibility index 2006.* Retrieved 14 July, 2006, from www.bitc.org.uk/crindex.

Coombs, H. C. (1970). *The fragile pattern: Institutions and man.* The 1970 Boyer Lectures. Canberra: Australian Broadcasting Commission.

Desailly, M., Bushell, R., Scott, J. E., Simmons, B. L., Sinha, C. C., & Baillie, B. (2004). Environmentally and socially responsible practices in the camping and caravan industry: A case study from Australia. *Tourism and Recreation Research, 29*(3), 39–50.

Global Reporting Initiative (GRI). (2002). *Sustainability reporting guidelines.* Amsterdam: Global Reporting Initiative.

Greene, D. (1999). *Carrots, sticks and useful tools: A study of measures to encourage cleaner production in Australia.* Sydney: CRC for Waste Management and Pollution Control, University of NSW.

Hamilton, C., & Denniss, R. (2005). *Affluenza: When too much is never enough.* Sydney: Allen & Unwin

Honey, M. (2002). *Ecotourism certification: Setting standards in practice.* Washington, DC: Island Press.

Lowe, I. (2005). Achieving a sustainable future. In J. Goldie, R. Douglas, & B. Furness (Eds.), *In search of sustainability.* Canberra: CSIRO Publishing.

NSW Council on Environmental Education. (2002). *Learning for sustainability, NSW environmental education plan 2002–2005. EPA 2002/77.* Sydney: New South Wales Government.

Simmons, B., Scott, J., Bushell, R., Sinha, C., Desailly, M., & Baillie, B. (2006). Environmental management partnerships: A research case study from the tourism industry. In R. Welford, P. Hills, & W. Young (Eds.), *Partnerships for sustainable development.* Hong Kong: The Centre of Urban Planning and Environmental Management, University of Hong Kong.

Smith, G., & Scott, J. (2006). *Living cities: An urban myth?* Sydney: Rosenberg Publishing.

UNEP. (1998). *Ecolabels in the tourism industry.* Paris: United Nations Environmental Programme.

UNEP. (2005). UN Decade of Education for Sustainable Development (2005–2014). Retrieved 14 July, 2006, from http://portal.unesco.org/education/en/ev.php. URL_ID=27234&URL_DO=DO_TOPIC&URL_SECTION=201.html.

Viere, T., Herzig, C., Schaltegger, S,. & Leung, R. (2006). Partnerships for sustainable business development: Capacity building in South East Asia. In R. Welford, P. Hills, & W. Young (Eds.), *Partnerships for sustainable development.* Hong Kong: The Centre of Urban Planning and Environmental Management, University of Hong Kong.

Chapter 11
Backcasting Using Principles for Implementing Cradle-to-Cradle

Freek van der Pluijm, Karen Marie Miller, and Augusto Cuginotti

Abstract This chapter explores the strategic implementation of the cradle-to-cradle concept and suggests backcasting using sustainability principles as a systematic way to support decision-makers. The cradle to cradle concept seeks to learn from nature and to design using principles that emphasize the conversion of waste into food, the use of solar energy inputs, and the celebration of diversity. As such, it facilitates organizational transition toward enabling a societal infrastructure by participating in cyclical supply chains – a valuable complement to the green supply chain approach to organizational collaboration for sustainability. The specific contribution of this chapter to the cradle-to-cradle literature focuses on the integration of cradle-to-cradle design within a systems approach that permits analysis from a strategic sustainable development perspective. After usefully comparing cradle-to-cradle design principles with FSSD principles for sustainability, the authors integrate science-based principles with value-based principles as an asset to support backcasting using overarching sustainability constraints drawn from scientific principles for socio-economic sustainability. This framework is one that decision-makers can use flexibly to make mid-course corrections in the march toward a societal infrastructure that supports a targeted system in which all material flows are either part of a biological or a technical metabolism.

Keywords Strategic sustainable development · Cradle-to-cradle · Systems approach backcasting · Sustainability principles · Decision-making

11.1 Introduction

In the current economic paradigm, growth is largely based on the deterioration of social and environmental systems. Linear models of production, such as the

F. van der Pluijm (✉)
Master's Programme "Strategic Leadership towards Sustainability", Blekinge Institute of Technology, Karlskrona, Sweden
e-mail: freek@strategicsustainabledevelopment.eu

J. Sarkis (eds.), *Facilitating Sustainable Innovation through Collaboration*,   203
DOI 10.1007/978-90-481-3159-4_11, © Springer Science+Business Media B.V. 2010

Take-Make-Waste model, externalize costs onto those with the weakest voice: society's poor and future generations. According to Meadows, Meadows, and Randers (1992), "without significant reductions in material and energy flows [into nature], there will be in the coming decades an uncontrollable decline". This linear way of interacting with nature is having cumulative and far reaching effects on the health of the biosphere. Not only are the effects of this accumulation of waste and degradation of natural systems accelerating, but the potential for redesign of the systems is also being undermined (Robèrt et al., 2004). The seemingly unrelated effects are interconnected and stem from the same underlying causes. Thus, addressing the root causes of the problems provides an opportunity to redesign issues out of the system at the source. The more we study the major problems of our time, the more we come to realize that they cannot be understood in isolation. These natural limits to growth need to be considered and understood in the development of new industrial processes. The biosphere works in cycles, and in order to interact with these systems in a sustainable way, the redesign of human society according to the paradigm of cyclical thinking is required.

One of the concepts used to inspire people to contribute to this transition is cradle-to-cradle. Cradle-to-cradle is build on the premise "learning from nature" and aims to design in such a way that "waste = food". In the Netherlands, cradle-to-cradle is gaining momentum. The shared language of cradle-to-cradle, with its focus on the creation of technical metabolisms, provides the unique potential to trigger conversations around the enabling of a societal infrastructure that supports organizations in their transition toward participation in cyclical supply chains. However, efforts that aim to contribute to cyclical production/consumption methods are difficult to implement. Society currently supports take-make-waste behaviour rather than supporting cyclical supply chains. Cradle-to-cradle gives direction to the needed societal transition toward cyclical relationships and its principles for design provide an appealing invitation for innovation in this realm.

In this paper, we suggest the integration of cradle-to-cradle design in a systems perspective, allowing it to be analyzed through the lens of a framework for strategic sustainable development in order to prioritize actions, allowing cradle-to-cradle to be a strategic stepping stone toward a sustainable society.

11.2 Cradle-to-Cradle

In a 1998 speech, William McDonough (1998), architect and co-author of the book Cradle to Cradle, describes the three defining characteristics that we can learn from natural design as follows:

1. Everything we have to work with is already here.

   - Everything is cycled constantly with all waste equalling food for other living systems.

2. Energy comes from outside the system in the form of perpetual solar income.

- It is an extraordinary complex and efficient system for creating and cycling nutrients, so economical that modern methods of manufacturing pale in comparison to the elegance of natural systems of production.

3. Biodiversity is the characteristic that sustains this complex and efficient system of metabolism and creation.

- What prevents living systems from running down and veering into chaos is miraculously intricate and symbiotic relationship between millions of organisms, no two of which are alike.

Based on this understanding, and on the understanding that society is inherently part of nature, humanity can design its systems for producing and living in accordance with this way of thinking. Concepts such as industrial ecology, biomimicry and cradle-to-cradle provide sets of simple design rules that allow us to learn from these characteristics of natural systems (Benyus, 2002; Ehrenfeld, 1997). From an industrial design perspective this means developing materials, products, supply chains, and manufacturing processes that replace industry's cradle-to-grave manufacturing model (McDonough & Braungart, 2002b).

The three tenets around which cradle-to-cradle is built are:

- Waste = Food
- Use current solar income
- Celebrate diversity

(McDonough & Braungart, 2002a)

Fig. 11.1 Graphical representation of the three key tenets of cradle-to-cradle: waste = food is represented by the (t)echnical and (b)iological metabolisms, use current solar income, celebrate diversity

The cradle-to-cradle approach specifically focuses on the concept of biological and technical metabolisms as a method to close material loops. In the biological metabolism, the nutrients that support life on Earth – water, oxygen, nitrogen, carbon dioxide – flow perpetually through regenerative cycles of growth, decay and rebirth in such a way that waste equals food (McDonough & Braungart, 2002a). The concept of cradle-to-cradle suggests that the technical metabolism can be designed to mirror natural nutrient cycles; as a closed-loop system in which valuable, high-tech synthetic materials and mineral resources circulate in an endless cycle of production, recovery and remanufacture (McDonough & Braungart, 2002a). In order to achieve a sustainable relationship between society and ecological systems, a societal infrastructure needs to be in place that enables the stream of materials to flow into either a biological metabolism or a technical metabolism.

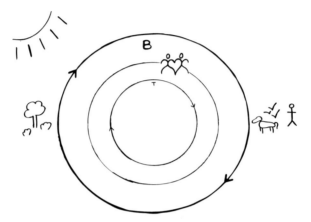

Fig. 11.2 Systems view of the components of cradle-to-cradle. The biological metabolism goes in cycles of photosynthesis and respiration and is driven by energy from the sun. The representation of the technical metabolism within the biological metabolism highlights the interconnectedness of the two metabolisms; the technical metabolism is created by society

This raises questions about to how to apply this concept in practice since the current economic system is set up in such a way where it is not necessarily economically viable to re-capture that waste and there is limited incentive to develop the societal infrastructure required for that transition. The discussion on an optimal way of encouraging the design of a societal infrastructure based on the cradle-to-cradle metabolisms is just beginning. In order to build this infrastructure, strategies need to be developed that support the transition toward this infrastructure, and tools need to be developed that support entrepreneurs and communities in making their contribution to the transition toward a cradle-to-cradle inspired sustainable infrastructure. Processes will need to be developed to complete the links for a circular supply chain. These new mechanisms will need to be designed to fit the needs of individual organizations, and collaboration and systems thinking will be key to ensuring that they also move in the direction of societal sustainability.

11.3 Current Status of Cradle-to-Cradle Implementation in the Netherlands

The cradle-to-cradle concept is, compared to the rest of the world, very popular in the Netherlands. After the broadcasting of a very compelling documentary on the concept in November 2006, many initiatives have begun in order to implement the concept. Through interviews with decision-makers working with cradle-to-cradle, a gap was identified between understanding of the system and the identification of actions and tools. In particular, there is a lack of shared understanding of success and a clear strategy for selecting actions and tools in line with a shared vision of success. In the context of the Netherlands, actions and tools are being selected and implemented in an ad-hoc way, and there is an uncertainty as to how to determine strategic steps to work toward cradle-to-cradle. There is a shared concern that unless clear strategies and concrete successes are achieved, this surge of enthusiasm around the cradle-to-cradle concept will fail to lead to real, and tangible, change toward sustainable development.

Current cradle-to-cradle implementation efforts lack a systems overview and strategic approach. Questions at the systems level to practitioners implementing cradle-to-cradle projects triggered responses that specifically related to their area of expertise. At the same time, responsibilities for certain key aspects with respect to the systemic implementation were shifted to other parties. For example, one interviewee stated "I trust that someone else will take care of the energy problem".

On the other hand, there is an enthusiasm and momentum behind the cradle-to-cradle concept, and people are energized to work in new and innovative ways to implement it. Research institutions and governments are devoting time and money toward developing the concept (Thesingh, Levels, Kersten, & Korevaar, 2008, personal communications). Networks are bringing together people from different sectors of society to interact in unconventional ways, with the shared intention of working toward sustainable development. The opportunity that arises from this shared enthusiasm for working toward sustainability based on the cradle-to-cradle concept is the possibility of having a constructive dialogue around building an infrastructure that supports the closing of material loops. Several players from all parts of the supply chain are involved and open to the exploration of possibilities to enable cradle-to-cradle production.

11.4 The Framework for Strategic Sustainable Development

In order to ensure the contribution of cradle-to-cradle inspired design to the transition toward a sustainable society, it is crucial to have a structured understanding of the systems to which we belong. All system levels are interconnected and interaction between the scales is such that change at one scale affects change at others (Holling, 2004).

The following five-level model can be used for planning in complex systems, where five hierarchically different system-levels are delineated. The distinction between the levels is maintained while planning and structuring information, while the interrelatedness between them is acknowledged and can then be utilized in a deliberate and methodical fashion. The five levels are (Robèrt et al., 2004, pp. 28–50):

1. **System**: Description of the *constitution* of the system and the dynamic interrelationship within and between the social and ecological systems.
2. **Success**: Definition of principles for a favorable *outcome* of planning within the system (e.g., *principles for sustainability and achievement of the vision*).
3. **Strategy**: Strategic *process* guidelines, build on backcasting, that inform a step-by-step approach toward success.
4. **Actions**: *i.e., concrete measures* that comply with the principles for the process to reach a favorable outcome in the system.
5. **Tools**: Any tool that enables the process of strategically working toward success. (*e.g., strategic tools: monitoring and reporting on process; systems tools: monitoring and reporting on the system; and capacity tools: tools that help people learn*).

The framework for strategic sustainable development[1] (FSSD) was developed through a process of scientific consensus at the principle level that has taken place in a learning dialogue between scientists and policy makers in business and politics. This process began in the mid 80s, and continues to evolve (Broman, Holmberg, & Robèrt, 2000). The framework is designed to provide strategic direction, a "compass" for organizations' sustainability initiatives by providing a generic framework within which information can be structured in a way that supports decision making. Such a framework, based on first order principles, allows decision makers to interpret details and understand strategies without losing sight of the bigger picture (Broman et al., 2000). This allows for improved effectiveness and strategic planning of actions in contributing to the process of sustainable development.

Within the generic five-level model, the FSSD approach to Sustainable Development defines the system based on the nested system model (Fig. 11.3). Specific principles for success and strategic guidelines form essential components of the FSSD, and are as follows:

Fig. 11.3 Nested system model

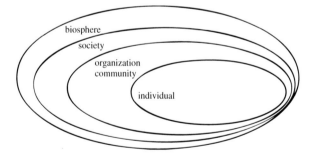

---

[1]This framework is published in many books, case studies and peer-reviewed journals within the field. It is often known amongst business leaders and policy-makers as The Natural Step Framework after the international non-profit organization that helped initiate its development and which continues to promote the strategic sustainable development approach.

11.4.1 Success

Based on study of the dynamic interrelationships between society and the biosphere, and an understanding of science; including thermodynamics and conservation laws, biogeochemical cycles, basic ecology, the primary production of photosynthesis; Dr. Karl-Henrik Robèrt initiated a process of collective scientific inquiry into the root causes of unsustainability. The following sustainability principles are a result of that work, and have been tested and refined over a period of approximately 20 years:

In a sustainable society, nature is not subject to systematically increasing...

  I. ...concentrations of substances extracted from the Earth's crust,
 II. ... concentrations of substances produced by society,
III. ... degradation by physical means.
      Social sustainability is addressed by the fourth sustainability principle:
      In a sustainable society...
 IV. ... people are not subject to conditions that systematically undermine their capacity to meet their needs.
      (Ny, MacDonald, Broman, Yamamoto, & Robèrt, 2006; Robèrt, 2000)

These sustainability principles have been specifically designed to support the strategic process of backcasting.

11.4.2 Strategy: Backcasting and Strategic Guidelines

Planning in complex systems is supported and guided by applying the concept of backcasting. Backcasting, as opposed to forecasting methods of predicting the future, is about working backward: setting the desired future state and looking back from the point of success to define which steps are needed to attain it. The main difference between the two is that backcasting focuses on designing how desirable futures can be attained and forecasting works on figuring out futures that are likely to happen (Robinson, 1990).

In the field of sustainability it is not sufficient to know scenarios of the future that are most likely to happen. Current sustainability problems are a result of the current trends and ways of thinking in society. Therefore, in order to strategically plan for the transformational change required to create a sustainable society, it is vital to plan normatively rather than by perpetuating current trends. Given multiple possible futures, decision makers are looking for the most desirable rather than the most likely outcome (Robinson, 1988).

The FSSD strategy focuses on the process of backcasting using sustainability principles, instead of focusing on the creation of a desired scenario. To accomplish this, the FSSD provides a set of sustainability principles to work as boundary conditions. As long as these principles are complied with, the scenario developed within these boundaries is more likely to represent a sustainable future. Backcasting using basic principles explicitly expresses the constraints of the system, and allows for creativity in the course of the development of strategy, actions, visions and goals while providing general rules to guide decisions in the right direction rather than providing a solidified or prescriptive vision of the future (Holmberg & Robèrt, 2000).

11.4.3 Actions and Tools

Any tools and actions that support strategies toward sustainable development are encouraged while applying the FSSD. These could range from community engagement and dialogue tools; such as community-based social marketing and world café dialogues, to technical tools such as energy monitoring systems and renewable energy technologies, to actions undertaken at a policy level, such as implementing a carbon tax. When various tools and concepts for sustainable development[2] are used within the five level model, their complimentary nature is highlighted, and it is easier to determine ways to use them in parallel, each for its specific purpose (Robèrt et al., 2002). Levels in the five-level model are interdependent, and diverse tools and actions are required at every level, selected according to context (Robèrt et al., 2002).

11.5 Supporting Cradle-to-Cradle with the Framework for Strategic Sustainable Development

There is a lot of potential for the integration of cradle-to-cradle and FSSD. Integrating sustainability principles within the cradle-to-cradle implementation strategy has the potential to ensure that the solutions designed are strategic steps toward sustainability at a systems level.

Although the systems view and the concept of closed loops are shared, cradle-to-cradle and the FSSD frame the creation of closed loop cycles in different ways in relation to the larger system. Cradle-to-cradle, with its focus on eco-effectiveness, seeks to redefine the concept of waste by framing wastes as nutrients that provide value to systems external to the boundaries of the system under consideration. In this way, while working toward eco-effectiveness, at times cradle-to-cradle even encourages the production of "wastes" because they produce value for another system. In this case, cradle-to-cradle communicates the opportunities for positive effects beyond the boundaries of the system, whereas the FSSD focuses on eliminating contributions that lead to the degradation of the suprasystem, i.e., society within the biosphere.

This distinction is subtle, but important, as cradle-to-cradle tends to focus on creating positive effects, and does not include clear criteria or guidelines to ensure

---

[2]Such as Natural Capitalism, Factor X, Ecological Footprinting, Life Cycle Assessment and Sustainable Technology Development.

systematic analysis that eliminates the creation of negative effects to the larger system. The FSSD, on the other hand, provides clear criteria based on the current scientific understanding of the natural systems and lends itself to the analysis of scale and equilibrium between the systems of society and the biosphere. Both the creation of opportunities and adherence to basic sustainability principles are crucial to sustainable development over the long-term, cradle-to-cradle and the FSSD cover both aspects.

Finally, searching for positive opportunity in design without a rigorous decision-making process leaves potential for problem displacement and problem shifting at the larger systems level. Cradle-to-cradle is based in systems understanding, and the FSSD provides the systematic approach to apply the concept in a strategic way with the larger purpose in mind.

11.6 Principles for Cradle-to-Cradle and FSSD

The cradle-to-cradle principles overlap substantially with the FSSD sustainability principles, and move in the same direction, as they are also principles for a sustainable society, grounded in an analysis of the same systems (Fig. 11.4).

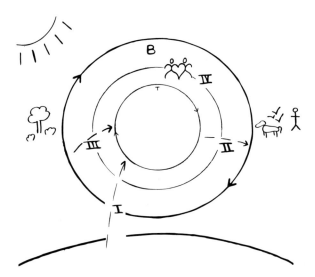

Fig. 11.4 Systems view including technical and biological loops: society within the biosphere, powered by energy from the sun. This figure also shows flows from the lithosphere and the biosphere to the technosphere, as well as from the technosphere to the biosphere. The roman numerals indicate the sustainability principle these flows are associated with; "waste = food" has the potential to lead to a reduced extraction flow in I, reduced degradation in III, and a reduced waste flow in II. "Use current solar income" has the potential to lead to a reduced flow in I. "Celebrate diversity" has the potential to lead to reduced violations in III and IV

Cradle-to-cradle Design Principles

- Waste = Food
- Use current solar income
- Celebrate diversity

FSSD Principles for Sustainability
In a sustainable society, nature is not subject to systematically increasing. . .

I. . . .concentrations of substances extracted from the Earth's crust,
II. . . . concentrations of substances produced by society,
III. . . . degradation by physical means.
  And in that society. . .
IV. . . . people are not subject to conditions that systematically undermine their capacity to meet their needs.
  (Ny et al., 2006; Robèrt, 2000)

The main difference identified, is that the FSSD principles are designed specifically for planning using backcasting, which means that they are framed in the negative in order to identify boundary conditions and have been scrutinized and improved to meet the following criteria as closely as possible: concrete, science-based, non-overlapping, general, necessary and sufficient (Robèrt et al., 2004). As such, they have a number of advantages in the context of decision-making for sustainability. They are sufficient and systematic, and analysis of decisions against the sustainability principles means that all aspects of sustainability are covered. In addition, they are concrete and general enough to be applied in any situation and to analyze specific decisions against.

The cradle-to-cradle principles, on the other hand, are framed in the positive in order to serve as sources of inspiration and are neither concrete nor systematic enough to guide specific decisions. However, they are appealing, easily understood and communicated, and they add color to the understanding of the system and offer an appealing description of success. They trigger creativity and provide inspiration for the design of specific scenarios with the potential to move toward compliance with the FSSD sustainability principles at a societal level, although they are not designed to scrutinize decisions against.

The following Table 11.1 shows a summary of characteristics of both sets of principles, highlighting the distinctions between the two ways of communicating principles of success.

Table 11.1 Summary comparison between cradle-to-cradle principles and sustainability principles

Cradle-to-cradle principles	Sustainability principles
Metaphor	Robust
Creative	Systematic
Design	Planning

11.7 Backcasting Using the Cradle-to-Cradle Concept

"It is important (...) that signals of intention be founded on healthy principles" (McDonough & Braungart, 2002a) in order to make sure that one problem is not being substituted for another. This approach is aligned with using principles for design, though, if these design principles are applied without a robust understanding of the system constraints, there is potential for the designed solutions to fall outside of the sustainability principles, and contribute to unsustainability over the long-term. On the other hand, if these design principles are applied in a way that is bounded by overarching sustainability constraints, then the power of the backcasting approach to planning for sustainability is maximized. The integration of both design principles and boundary conditions shows strong potential as a robust planning process for transformational change.

This planning methodology, backcasting using sustainability principles, is proposed to overcome the challenge of strategic implementation of cradle-to-cradle. The constraints for the future desired state are set by the four scientific principles for socio-ecological sustainability. Within these basic scientific constraints creativity is allowed and encouraged, allowing the space for inspiration and true innovation based on cradle-to-cradle thinking and keeping in place the systems perspective to ensure that the chosen actions are strategic stepping stones toward a more sustainable society. Figure 11.5 represents the generic steps for implementation through backcasting.

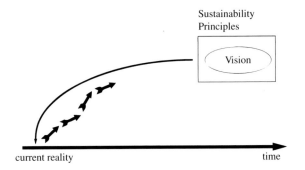

Fig. 11.5 Process of backcasting from a designed success state

I. Designing principles of a future state.

The definition of success for the society or organization are identified in the first stage through the discovery of core purpose, core values and existing strategic goals. The principles for sustainability act as the boundaries of the "design space" within the envisioned sustainable society. The concept of cradle-to-cradle offers inspiring guidelines and metaphors for the design of products and processes for sustainability. Together with social principles derived from stakeholder engagement, the success principles for a society, project or organization are designed.

II. Analysis of current reality.

By looking at the current reality through the lens of the sustainability principles and the vision of success, an analysis takes place of the gap between the vision of success and the current reality.

III. Creating compelling measures.

The identified gap serves as a creative tension based on which actions and measures are brainstormed and identified that have the potential to positively contribute to a shift toward the vision.

IV. Setting priorities.

The proposed measures are scrutinized against, at a minimum, the following three prioritization questions and a cross-check takes place between the final list and the gap-analysis of step II;

- Is this measure a step in the right direction with regards to the sustainability principles and the organization's vision of success?
- Is this measure a flexible platform for future development toward the vision, and full compliance with the sustainability principles?
- Does this measure provide sufficient return on investment to continue the process?

After these four steps have been followed, a strategic action plan can be developed to implement the suggested measures and start strategically planning in the direction of a more sustainable society (Robèrt et al., 2004, pp. 242–248). The specific content of a strategic action plan created will differ significantly depending on the context of the organization applying the process. One example is the Canadian municipality of Whistler, where the "backcasting using sustainability principles" planning method was applied to design their award-winning[3] integrated sustainability plan. The participatory process based on the above framework, allowed room for creativity and community involvement, applied in a robust and systemic way. The resulting plan had 16 strategic focus areas, ranging from Art, Culture and Heritage, to Natural Areas and Resident Housing. The final comprehensive plan included thousands of actions, providing a comprehensive plan for sustainable development in the region (Waldron, 2008).

11.8 Synergetic Nature of Cradle-to-Cradle and the FSSD

The first stage of a paradigm shift is metaphoric and the second stage is descriptive and analytic. In order for a paradigm shift to occur, a deep change is needed in the first stage, not only a change in the second stage (Ehrenfeld, 1997; Korhonen, 2002).

---

[3]Whistler was awarded the "best long term planning" award at the 2005 United Nations Liveable Communities Awards in LaCarna, Spain.

The FSSD has a structured planning and decision-making process to implement specific actions toward this new paradigm of cyclical relationships in a systematic way, and cradle-to-cradle has the potential to engage and communicate this new paradigm to many people.

The FSSD is an inclusive approach to sustainable development, which is based upon backcasting from a desired future, structuring information in a systematic way to enable decision-making, and designed for the incorporation of diverse tools and concepts that support the strategic goals. This provides an open and flexible approach that supports sustainable development initiatives.

Cradle-to-cradle supports sustainable development by looking beyond the minimum requirements for survival and searching for ways to create opportunities. It holds a vision of human industry as a regenerative force and searches for ways to restore nature and create enduring wealth and social value (McDonough & Braungart, 2002b). The tools and strategies applied are designed to trigger creativity, which spurs innovation, and are based upon principles of success from the powerful positive metaphor "learning from nature". This concept has the power to inspire individuals, and to spark conversations and collaboration among diverse members of society, as we have seen in the Netherlands.

For managers and policy makers, this inspiration and innovation is best harnessed when coupled with a systemic and strategic framework for decision making. The FSSD provides this systematic approach, and has been designed to provide a shared language to facilitate collaboration for sustainable development. Combining the inspiring vision, offered by cradle-to-cradle, with the robust principled planning approach offered by the FSSD provides a strong foundation for collaboration and innovation for sustainable development.

Acknowledgments We would like to acknowledge the support of our advisors, HenrikNy and Pong Leung, classmates, the Master's in Strategic Leadership toward Sustainability (MSLS) program team, Dr. Karl-Henrik Robèrt and all our interviewees, both while visiting the Netherlands, in person and through conference calls.

A special remark goes to the cooperation of Enviu – Innovators in sustainability! represented by the enthusiastic support of Wouter Kersten. We would finally like to thank representatives from the Province of Limburg, especially for the energy of Frederieke Vriends and the commitment of Paul Levels and Joey Clark.

References

Benyus, J. M. (2002). *Biomimicry: Innovation inspired by nature*. New York: Perennial.

Broman, G., Holmberg, J., & Robèrt, K. H. (2000). Simplicity without reduction: Thinking upstream towards the sustainable society. *Interfaces: International Journal of the Institute for Operations Research and the Management Sciences, 30*(3), 13–25.

Ehrenfeld, J. R. (1997). Industrial ecology: A framework for product and process design. *Journal of Cleaner Production, 5*(1–2), 87–95.

Holling, C. S. (2004). From complex regions to complex worlds. *Ecology and Society, 9*(1):11. [Online] URL: http://www.ecologyandsociety.org/vol9/iss1/art11

Holmberg, J., & Robert. K. H. (2000). Backcasting from non-overlapping sustainability principles a framework for strategic planning. *International Journal of Sustainable Development and World Ecology, 7*(4), 291–308.

Korhonen, J. (2002). The dominant economics paradigm and corporate social responsibility. *Corporate Social Responsibility and Environmental Management, 9*, 67–80.

McDonough, W. (1998). Essay: A centennial sermon: Design, ecology, ethics, and the making of things. *Perspecta, 29*, 78–85.

McDonough, W., & Braungart, M. (2002a). Cradle to cradle: Remaking the way we make things. New York: North Point Press.

McDonough, W., & Braungart, M. (2002b). Design for the triple top line: New tools for sustainable commerce. *Corporate Environmental Strategy, 9*(3), 251–258.

Meadows, D. H., Meadows, D. L., & Randers, J. (1992). *Beyond the limits: Confronting global collapse, envisioning a sustainable future*. Post Mills, VT: Chelsea Green.

Ny, H., MacDonald, J., Broman, G., Yamamoto, R., & Robèrt, K.-H. (2006). Sustainability constraints as system boundaries an approach to making life-cycle management strategic. *Journal of Industrial Ecology, 10*(1–2), 61–77.

Robèrt, K. H. (2000). Tools and concepts for sustainable development, how do they relate to a general framework for sustainable development and to each other. *Journal of Cleaner Production 8*(3), 243–254.

Robèrt, K. H., Broman, G., Waldron, D., Ny, H., Hallstedt, S., Cook, D., et al. (2004). *Strategic leadership towards sustainability*. Karlskrona: BlekingeTekniska Högskola.

Robèrt, K. H., Schmidt-Bleek, B., Aloisi de Larderel, J., Basile, G., Jansen, J. L., Kuehr, R., et al. (2002). Strategic sustainable development selection, design and synergies of applied tools. *Journal of Cleaner Production, 10*, 197–214.

Robinson, J. B. (1988). Unlearning and backcasting: Rethinking some of the questions we ask about the future. *Technological Forecasting and Social Change, 33*, 325–338.

Robinson, J. B. (1990). Futures under glass: A recipe for people who hate to predict. *Futures, 22*(8), 820–841.

Waldron, D. (2008). Applications of the FSSD. Lecture slides from a presentation to graduate students at the BlekingeTekniska Högskola, December 9, 2008, Karlskrona, Sweden.

Chapter 12
Corporate Strategies for Sustainable Innovations

Marlen Arnold

Abstract Rerouting social and corporate activities concerning sustainable change is one of the key challenges for many businesses today and in the near future. Implementing sustainable requirements in corporations necessitates the initiation of corporate strategic change and the development of sustainable innovations. In the light of a wide variation of corporate activities to cope with these challenges, the following questions arise: (1) When and why do companies pursue processes of strategic change to integrate sustainability and develop innovations? (2) Which effects and extents do these sustainable values and innovations have, and (3) what factors promote or inhibit sustainable strategic change? This study highlights integrative strategies for sustainable innovations on the basis of a case analysis of three companies and examines the organisational, the cultural as well as the structural conditions for active sustainable oriented corporate policies. The study analyses internal and external explanatory factors for the occurrence of sustainable strategic change processes, the conditions for a company's commitment to sustainability, the conditions that result in strategic change, and the capability to generate sustainability-oriented (product) innovations in medium-sized and large companies. Moreover, it develops a framework that integrates both aspects of strategic content and strategy formation processes regarding sustainability. Thus, along with the strategic change literature the study distinguishes between the timing (i.e. proactive or reactive change) and the intensity of strategic change regarding sustainability and innovation strategies. The findings highlight the role of visions and options, the company interactions, the role of change agents and management and their values and norms, the companies' history, and the history of business fields. It can be shown that proactive companies and companies having a high level of sustainable impact in their strategic changes or innovations depend on different influencing factors.

M. Arnold (✉)
TUM Business School, Technische Universität München, Alte Akademie 14, Freising, 85350, Germany
e-mail: marlen.arnold@wi.tum.de

Keywords Corporate strategic change · Change agents · Sustainable values and norms · Sustainability strategies · Innovation strategies

12.1 Introduction

Innovations are generally anchored as patterns in our society. Companies are forced to innovate continuously, to change the corresponding strategy and to constitute an adequate strategic flexibility. Marketing and non-marketing strategic changes and innovations also are of striking significance for the successful move towards a sustainable path; on the other hand, they can cause the creative destruction of established and/or sustainable economic ways, needs and social constructions (Erdmann, 1993, Sauer, 1999). The consequences, range and interference level of innovations are difficult to estimate ex-ante and only reveal themselves gradually through their use (Sauer, 1999). The main problem concerns the development of relevant knowledge to solve problems before new problems emerge. In this regard, technological, social and organisational contexts are closely interlaced (Rohracher, 1999) within a strategic management context.

The focus of this study is on the analysis of sustainability-oriented change processes in terms of supportive and limiting factors. Three case studies are utilised to show the interdependency between external and entrepreneurial factors as well as the connection between strategic contents and strategic formation processes. Novamont, a Swedish small-sized chemical firm, Bedminster, an Italian medium-sized recycler, and Philips, a large Dutch electronics company were used to illustrate sustainability-oriented patterns.

12.2 Theoretical Background

Sustainability is now considered a relevant issue in strategic management (Hammer, 1998; Hill & Jones, 2008), and corporate strategic management continues to grapple with the challenges and ideas of sustainable development (Aragón-Correa & Matias-Reche, 2005; Hart, 1995; Porter & van der Linde, 1995; Reinhardt, 1998).

In strategic management, the structural and material action's margin of a venture and its limitation can be approached through history, path dependency, actual resources and environment constellation. Those aspects offer a good frame of reference to clarify the possibilities and limits of companies for initiating sustainable strategic change.

Corporate cultural elements also play a role in sustainable strategy formation and help identify the options for sustainability-oriented intervention. Corporate culture can be understood as the congruent interplay of all persons, relations and elements in processes and appears in relations as well as in the human behaviour and in the form of symbols (Alvesson, 2005). Corporate culture conveys both the value of a company and the expression of the established solutions and social integration, and thus plays a key role in the strategy formation processes. According to Schmidt, corporate culture ensures first and foremost 'the identity, the efficiency, the dynamic and the crisis competence of a company and constitutes therewith in principle firstly its marketability and trademark capability' (Schmidt, 2005, p. 112). The success of these capabilities is closely linked to the organisation of entrepreneurship and the members' freedom in a company.

Sustainability-oriented strategic change embraces the strategy content as well as the view on strategy formation process. Both resource and market based factors are relevant for strategic management (Hill & Jones, 2008). Accordingly, organisational, cultural and external factors are the three main issues to explain sustainability-oriented strategic change in this study.

According to Dutton and Duncan (1987) the interest in strategic change arises when the company perceives its strategic fit as inadequate. Then, the perception of urgency and the capability for strategic change play a significant role in the change process on the basis of the assumed continuity of environmental changes and the visibility of these changes in the stakeholders' perception. The pressure to change the strategy is higher if change is long-term. Management's cognitive frames and their strategic understanding of the decision making process are also essential, along with resources, capabilities and the permanence of environmental changes (Burmann, 2001). Strategic change can also be proactive or reactive. Proactive change occurs early whereas reactive change has the tendency to take place later (Brockhoff, 1999; Brockhoff & Leker, 1998; Meyer, Goes, & Brooks, 1995; Nadler, 1988; Rajagopalan & Spreitzer, 1996; Zajac et al., 2000; Burmann, 2002; Lenker, 2000; Evans, 1991; Lambrechts, 2008; Shankar, 2009).

The goal of this empirical study is to investigate how sustainability-oriented changes were initiated and carried out in the three focal organisations. Influencing factors that seem to be causal for strategic change processes will also be analysed. In order to identify the conditions for the emergence of proactive-early strategic formation processes (e.g. early activities, willingness to take risks) and those containing a high sustainability impact (new or far-reaching processes, products and services, new kinds of satisfying needs), a conceptual framework was developed (see Fig. 12.1). The conceptual framework contains the three main influencing factors discussed above and related items.

In strategy formation, changes in (realised) achievement patterns are essential in terms of a company's fields of activity, entrepreneurial positioning, and resources. These assist with the maintenance, deployment and the development of organisational capabilities (Brockhoff, 1999; Brockhoff & Leker, 1998; Rajagopalan & Spreitzer, 1996). The question is how far and to what extent new values, aims, resources, etc. are going to be followed and applied by a company. Concerning sustainable development, the relevant question is how far and to what extent sustainability-oriented values and aims are aligned with the target system of a company and could or would show up in corresponding entrepreneurial activities. This means not only that the extent of the changes is relevant, but also the way the company integrates sustainable oriented content in its strategy formation process. Thus,

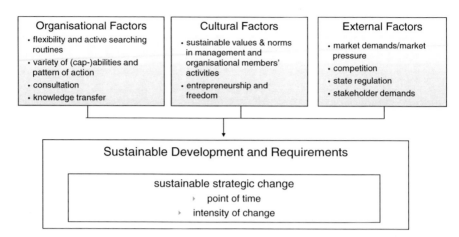

Fig. 12.1 Conceptual framework

in the context of this article, sustainability-oriented strategic change is defined as new (realised) processes and patterns, which show a reference to a company's field of activity (entrepreneurial positioning, resources, maintenance and development of organisational capacity) and have a clear environmental and/or social improvement.

12.3 Research Method

Methodologically, the study seeks to find causal relations based on the proposed framework. A qualitative multiple case study design was chosen, as this allowed research on complex social topics, such as innovation processes and strategic change with a focus on sustainability. The field studies of the three focal companies were carried out by semi-structured and thematically focused interviews, and supported by desk-top studies of related documents (Mayring, 2002; Yin, 1994). Data collection through semi-structured and theme centred interviews supported by document analyses was helpful in acquiring deeper insights into regions which are hard to access, such as sustainability- oriented strategic changes, and also to comprehend the change of sustainability demands and identify relevant influencing factors (Mayring, 2002). The data were collected orally in order to catch subjective meaning and sense attribution, which were very relevant in relation to sustainability and strategic formation processes (Yin, 1994). Among the central contexts were:

- The description of a company's internal sustainable development and the shaping and anchoring of sustainability into the company and its daily and ongoing routines.
- Organisational and personal characteristics as well as the option of official and unofficial exchange.
- Sustainable oriented cooperation patterns and structures.

The study's qualitative design aimed at both subjective perceptions and attributions of the individuals studied and also observable objective changes. Case selection was based on the companies' demonstrated results in corporate sustainable management. Thus, Novamont, Bedminster and Philips were chosen due to the implementation of environmental or sustainability relevant innovations, the fact that the innovations in question had a high strategic relevance for the company and the active generation of sustainable future markets, as well as the high sustainability-oriented intensity and range of the implemented innovative concepts. In addition, these three companies had experience in the interaction with stakeholders. The changes in question also had to be successful for at least 5 years to validate a new strategy.

Novamont develops biodegradable plastics to increase environmental and sustainable options, especially. While still a part of the Italian chemical company Montedison in 1989, Novamont started its research with the idea of linking chemistry and nature, investing millions as an independent company in the development of biodegradable materials. The recycling company Bedminster, formerly located in Sweden, converted a concrete factory in 1996 into a composting facility in Stora Vika near Stockholm to compost the city waste until 2003 which was then sold as refined organic fertiliser. Philips implemented a sustainability program. Using life-cycle analysis Philips developed six so-called Green Focal Areas 'weight, energy, packaging, recycling and disposal, lifetime reliability and hazardous substances' by which the products are evaluated. To be considered a Green Flagship, a product undergoes a divisional eco-design procedure, and is then investigated in the Green Focal Areas. Compared to a competitive product or forerunner a product must be proven to offer at least 10% improved environmental performance in at least one Green Focal Area and an overall lifecycle score that is equal or better.

From May 2002 through October 2005, a total of 15 persons from research and development departments, from sustainability or environmental units and from general management were interviewed. The entire study was built on semi-structured personal or telephone interviews (30–90 min). In most cases, an additional written survey was conducted to address follow-up questions. These surveys were necessary to collect more detailed or missing data, especially to have a comparable database of different companies or to check given data.

Content analysis was used to interpret the data (Yin, 1994). Data analysis used a coding system according to the analytical framework; each of the factors was operationalised by several codes (Mayring, 2002, 2003). In order to identify the conditions for the emergence of sustainable strategic change processes, a code system of cause-effect combinations was developed. After analysing each case study, all companies' results were aggregated on the basis of organisational, cultural, and external factors and compared with each other. The main aim was to find cross-company patterns. The combination of intensive individual case analysis and multiple case studies made it possible to systemise data based on individual cases and to gain a deeper insight into context and strategic changes (Eisenhardt, 1989; Gassmann, 1999; Yin, 1994). Furthermore, period oriented changes regarding historical and future perspective were included. Neither studies, interviews, nor questionnaires were limited to current strategies. Instead they were based on past decisions and patterns of action as well as on the estimation of future visions and activities.

12.4 Comparative Findings

Each company interprets the frame of the sustainable development concept and incorporates sustainability into the company using its own approach. All three strategies are proactive strategies in terms of environmental protection, maintenance of human and economic life and the focus on life cycle aspects. At Novamont the strategic renewal is based on the principle of the circular flow economy and its pioneer role is located in the area of using renewable resources. Bedminster, in contrast, focuses on zero emission concepts. At Philips, the life cycle concept, reinforced in the production concepts, was integrated with the Eco Vision-program as a specific sustainability program to mitigate the far-reaching effects of anthropogenic activities. Table 12.1 outlines the supporting factors for the respective strategic change.

Sustainability-oriented strategic change cannot be described by unique dominant influencing factors. Instead it is a dynamic interaction of diverse effects (Cleff & Rennings, 1999; Fichter & Arnold, 2005; Nagel & Wimmer, 2002). In the following, the major case study results are pointed out and compared while highlighting especially the key influencing factors for the early-proactive strategic change.

Table 12.1 Sustainability oriented changes in the companies

	Supporting factors for strategic change	Strategic change
Philips	– ISO 14001-certification of all units – Eco-balancing in the whole company – Institutionalisation of a board of sustainability – Implementation of sustainability division – Adoption of annual and biannual NH-Reports – Integration of the works council regarding sustainability concepts – Stakeholder integration in environmental themes – Changes in construction towards LCA – Consideration of 'using phase' in product development – Establishing of concrete sustainability cooperation – Integration of external consultants in sustainability projects – Institutionalisation of Intranet/Platforms/Network for the exchange of sustainability related information	Eco Vision-Program→ Six Green Focal Areas→Implementation of Green Flagships (Ecological pioneer products)

Table 12.1 (continued)

	Supporting factors for strategic change	Strategic change
Novamont	– Spin-off of the Italian chemical concern Montedison – Long-term research in the area of chemistry – The vision to combine chemistry and nature – Successful patenting of bioplastics in terms of circular flow economy – Establishing of concrete sustainability cooperation and distribution – Best practice projects /research projects in Germany – Supportive state regulation in Italy – European norm standards, i.e. CEN 13432 – European working group 'Renewable resources' within the program on climate change – High sustainability orientation of the management	Circular flow economy: variety of biodegradable material with the brand name Mater-Bi
Bedminster	– Joint venture of two technology companies – Existing infrastructure: conversion of an old cement factory into an in-vessel composting system – Experience in the area of mass composting facilities – Development and transfer of licences for waste treatment and recycling technology – Close cooperation with universities, public authorities and fast food companies – European waste framework directive 75/442/EWG – High sustainability orientation and technical know-how of the management – Cooperation with ZERI (Zero Emission Research & Initiatives)	Zero emission strategy in waste management/ composting, especially up-cycling as system-innovation

12.5 Organisational Factors

Flexibility: All three companies were proactively searching for sustainability solutions. The early search for sustainable alternatives is reflected in the current potential for action. Novamont's success is based on long-term cooperation with business partners. Thus, Novamont expands its distribution cooperation as needed. Even though Bedminster was searching for sustainable solutions and methods at an early stage, the market was not yet ready and the organisational capacities proved to be too weak in preventing insolvency.

Variety: The complex organisational capabilities of companies were revealed in the analyses, especially in cooperation, strategic alliances, monitoring and in the ability for dialogue. They have influenced the respective strategic changes significantly. Philips, in particular, possesses regular business as well as specific sustainability-oriented alliances and cooperation and thus shows a highly sustainability-oriented capability of development. Philips transforms cooperation in even more resourceful sustainable arrangements and uses them for further sustainable oriented strategic changes.

Consultation: The importance of advisors, process consultants or change agents for sustainable development was clear in all three company studies. In the case of Bedminster it was possible to speed up and support sustainable oriented strategic changes through intensive professional and procedural coaching by externs (sustainability experts, agents of governmental institutions etc.). In the case of Novamont the intensive strategic change was especially benefited from cooperation and strategic networks, whereas in the case of Philips consultants encouraged rather evolutionary sustainable steps. Even though the influence of externs on the intensity of strategic change varied, external consultants were always useful in promoting the integration of sustainable oriented knowledge in companies.

Knowledge: Group- and project oriented work as well as platform oriented exchange preserved the innovative potential of the employees and facilitated various improvements, especially in the case of Philips and Novamont (Becke, 2003). The platform structure of Philips also facilitated an effective exchange of knowledge since the experts communicate with each other directly (Howaldt, Klatt, & Kopp, 2004).

12.6 Cultural Factors

Values and members: A sustainability-oriented management is a crucial aspect of organisational culture. In the case of Bedminster and Novamont such strategic changes would have been impossible without the driving force of sustainability-oriented entrepreneurial vision and management (Bonsen, 2000; Senge, 2002). In addition, the sustainable vision targeted change requiring the support of the organisations' members (Klimecki, 1997). The organisational capabilities can be well connected at this point. Both SMCs showed that sustainability-oriented strategic changes are closely linked to the management perception of problems. All three companies are characterised by a clear sustainable vision which is supported and followed by strategic management (Galavan, 2008). At Philips the formation of partial strategies was highly reflexive and built on existing organisational capabilities. Bedminster exhibited weaknesses in management and leadership that had an impact on the entrepreneurial success. In the case of Novamont it is very explicit that the successful strategic change is conditioned by the strict focussing on sustainability-oriented targets and especially on the company's intention to change towards more sustainability. Philips and Novamont possess strong references to their foundation period and to formerly established values, norms and basic assumptions (e.g. greening the world, social responsibility for organisational members). Both companies benefit from the founders' appraisal and their philosophy – important elements for a successful strategic change.

Freedom: Entrepreneurship, cooperation and the creation of networks as well as the communication and the transfer of knowledge within an organisation are essential factors to support a promising sustainable strategic change. The capability to push sustainability-oriented strategic change forward with public authorities, private companies and non-governmental organisation in the form of public-private partnerships is also important. It follows Pfriem's (2005) view implying the increasing socialisation of the innovation process today via entrepreneurial embedment in communication, cooperation and networks. Sustainability-oriented knowledge could rapidly and efficiently be anchored in companies and establish free spaces for entrepreneurship on the basis of a dialogue oriented leading style, platforms, project and team work, (Argyris & Schön, 1996). Those free spaces count as elementary requirements for a functioning (i.e. conveying learning processes) knowledge management of open structure, transparency over goals, strategy and projects and the overlapping hierarchical communication processes (Minder, 2001).

Those free spaces conveyed evolutionary steps towards sustainability at Philips even though the company has a strong law orientation and a top-down orientation in the area of sustainability. These aspects had considerably carried strategic change at Bedminster as well. The internal change agents could be identified primarily in both SMCs with persons in management position or in interfaces of the F&E-division, while in the large scale company the prior initiative came from the sustainability division (Pfriem & Schwarzer, 2004). This also coincides with the recognition of Sharma & Vredenburg (1998, p. 741) that 'Solutions for reducing environmental impacts were often left to discretion of line managers. This discretion was accompanied by the integration of knowledge acquired from stakeholders, diffusion of knowledge within the organisation, keeping up the momentum of learning, and feedback of knowledge application.'

12.7 External Factors

Market: The identification of market opportunities demands is of great importance. In all three companies the early initiation of strategic change was determined or motivated by a high sustainability orientation of the market (Teece, Pisano, & Shuen, 1997).

Competition: Furthermore, market pressure or the entrepreneurial activities of the competitors initiate sustainability-oriented strategic changes by provoking reflecting and searching processes in the company which are the sources for the development of new solutions (March, 1991; Walgenbach, 2000). However, the condition is that the organisation recognises changes in its relevant field and implements (or is able to implement) them into real entrepreneurial activities. At the same time the searching and reflection processes require a tight interconnection with ideas and guiding principles of a sustainable development, like zero emission or life-cycle concepts to secure strategic change towards sustainability. The entrepreneurial activities of

Novamont and Philips on the market are joined with those of competitors. The pioneer activities of Novamont initiated other companies' activities in the market of biodegradable plastic production; in the case of Philips the ecological product line was matched with the products of its competitors.

State: State interventions, like laws, guidelines, supportive and research measures are important influential factors for proactive strategic change (Beckenbach & Nill, 2005; Porter & van der Linde, 1995; Siebenhüner & Müller, 2003). The intensity of the sustainability-oriented strategic change depends less on state initiatives than on organisational capabilities (e.g. resources, the history of the company), the strategic understanding of the management and current activities in strategic fields. State impulses can either operate as an initiator or ease the birth of new ideas or operate merely as a guide on the way to sustainability-oriented strategic change. In this regard it is crucial how strong the legal requirements and norms for the adoption of a procedure or the implementation of a strategy are. In those examples the state interventions contribute less to a creation of awareness and an initiation of strategic change and have less influence on the strategy intensity as well, but rather much more on the moment of action within a company. Sustainability relevant legal requirement changes offer essential indication on the road to strategic change by constantly adjusting the basic requirements. Legal requirements, market opportunities, corporate vision and sustainable entrepreneurship were the driving forces at both Novamont and Bedminster. Furthermore, state interventions convey pro-active strategic change by acting as an 'obstetrician' for sustainable technology, business concepts or strategy in the relevant product and service markets. Therefore, basic intervention openness is necessary as Philips and Novamont showed. At Philips, the state interventions mostly played the role of an initiator for the implementation of a partial sustainable strategy. Specifically, the new regulation of the EC in the area of electronic utilisation helped considerably creating new and stronger sustainability-oriented product concepts.

Stakeholders: Stakeholder requirements are a further influencing factor of sustainability-oriented strategic change in combination with other factors (Dyllick, 1989; Hedberg, 1981). Stakeholders had an influence on the strategic change in both SMCs and the large scale company. In this respect, the size of the company did not have a significant difference. A difference, however, can be shown in the intensity of the strategic change and the kind of entrepreneurial motivation taking those requirements into account. At Bedminster, the stakeholders had a firm and supporting function in the strategic change process since in this case the entrepreneurial interest and the stakeholders' interests were close. The tight collaboration with the stakeholders made it possible to increase the intensity of strategic change as well. Philips, on the other hand, picked up actual or potential stakeholder requirements, thus maintaining a good image and retaining the ability to be competitive. In this case the different business branches and the positioning of the respective or trendsetting companies on the market are significant factors.

Figure 12.2 shows the relationship between the main influencing factors regarding proactive strategic change and a high intensity of strategic change with respect to sustainability.

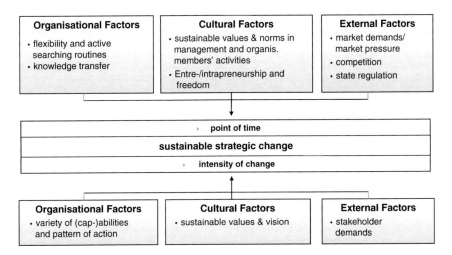

Fig. 12.2 Results regarding conceptual framework

12.8 Discussion

As can be seen from the examples of the three companies sustainability is definitively a strategic factor. Sustainability aligned strategic change processes lead to success or at least can be used as a learning model. In addition, the case study made clear that the companies had different opportunities and potential for taking up sustainable requirements and implementing concrete sustainable actions (Beckenbach & Nill, 2005).

Beside entrepreneurial interest the strategic change at Bedminster and Novamont had been initiated to take on increasing social needs and to realise them in offers proactively. Due to their activities all three companies designed new structures of sustainable economy or created sustainable structure proactively. A pattern change towards dialogue orientation and a product oriented exchange as well as project work to assimilate and implement sustainable requirements rapidly and efficiently into the company are exemplified in the case of Philips. Philips has developed a green product line and connects therewith to given structures and products. The strategic change contributes to sustainability contributions with regard to efficiency, prevention and consistence, but remains in a niche. Although Philips exhibits an early point of action in its new strategic alignment, the intensity of the strategic change is lower in comparison to Novamont and Bedminster. This is substantiated by the fact that the innovations offered by both SMCs possess a higher technological distance to the previous entrepreneurial activities. At Bedminster, the distance to the market was a crucial factor and made the enduring success impossible. Additional factors for the final failure were: (1) the insufficient amount of utilisable waste, especially mixed-waste, (2) difficulties in behavioural changes of consumers concerning waste disposal, (3) instable cooperation or difficult and complex forms of contract with partners and ministerial officers, (4) insufficient (existing) infrastructure – transport of fertilisers by boat, and (5) marketing problems due to mental barrier: 'fertiliser made from waste is bad'.

The success of Novamont's proactive and intensive strategic change is caused by marketability and the diversity and compatibility of its organisational capability.[1] Moreover, being a spin-off of an organically grown company having close cooperation and valuable strategic partners and being able to bear entrepreneurial risks enabled Novamont to achieve successful strategic change. For this reason Novamont was able to bear the technological risk and benefit from the resulting market chances. Even though Bedminster had dynamic and transferable capabilities, it had to develop its organisational capability more vigorously than Novamont and Philips to cover the technological and the market distance.

An additional explanation for the higher intensity of strategic change of the SMCs is offered by the companies' size. Reasons for Philips having no comprehensive strategic reorientation and adapting small relevant parameters instead are the company's market power, market positioning, and organisational inertia. The historical development and the current phase of Philips' cultural development convey rather less intensive strategic change of the company. According to Schein, (2003, 2008), in this phase a new cultural orientation, especially a successful cultural change regarding the differentiation and the obtainment of competitive advantage are necessary.

Finally, further research is necessary due to the chosen research design. The influence of investors regarding strategy formation and the strategic change were not considered in this case study design. As those stakeholders also affect the emergence and implementation of sustainability-oriented strategic change by the mean of providing or suspending financial capital, it would be interesting to work out the influencing modalities of investors, stockholders, banks or insurance firms. The structural factors, e.g. the size of the company, the organisational structure, the infrastructure, the logistic, etc. were not explicitly examined in the present study. For this reason, further investigations are required.

12.9 Recommendations and Conclusion

All in all, various patterns are recognisable for sustainability-oriented strategic changes (Beckenbach & Nill, 2005). However, an interaction concerning the need for sustainable products and services, sustainable oriented entrepreneurship and visions as well as distinctive organisational capabilities, such as the realistic estimation of the market or market appraisal and the development of necessary networks and cooperation, become apparent. Companies initially apply sustainable oriented

---

[1] According to Beckenbach and Nill (2005, p. 74) biodegradable materials can be seen as conversion technologies or hybrid solutions, which minimise the barrier between technologies as well as the transformation costs, serving therewith as lock-in loosener.

requirements as parallel strategies or in parallel paths. Market demands as well as governmental interventions influence strategic decisions, i.e. if those 'sustainability projects' are hived or substitute existing products and services (Arnold, 2007). Organisational size does appear to have an influence on sustainable strategic change, especially regarding the point of time and the intensity of change. If smaller companies decide to adopt a leadership position, a higher intensity of sustainable strategic change may well be possible. Therefore, a clear vision and management support are necessary. Companies that want to become more sustainable would benefit from education, collaboration and 'green' employees. Organisational change agents are of vital importance in strategic change processes. Externally, as shown, entrepreneurial activities are closely linked to the competitors in the same industry. Hence, big players will have the power to change market structures towards more sustainability – if they want to have a sustainable leadership position.

The available resources of a company and organisational capabilities as well as the learning ability and learning disposition of the company members determine the pace of organisational development and the intensity of the strategic change. From the point of view of an already established company generating completely new markets requires a particular motivation and high learning capability of the organisational members. Thus, formal learning processes and structures, such as networks, platforms, projects and workshops to transfer knowledge, are required. External consultants and change agents can stabilise the strategic change due to their profession and process knowledge or accompany, lead or even initiate change phases; yet, they cannot determine those strategic changes (Fichter & Arnold, 2005). Beside those factors structural issues (infrastructure, logistic, factory's location) have a fundamental significance for the success of an intensive strategic change and sustainability-oriented pioneer work in the case of Bedminster. Here, companies should invest time and money to strengthen the company's new strategy or to stabilise new segments, e.g. by cooperation, alliances, spin-offs, merging. Companies should use market chances to innovate sustainably, but if a company has no further possibilities of collaboration and the distance to the market is too great, it should think of taking other opportunities to innovate sustainably. For example, integration consumers or stakeholders in an early phase of sustainable strategic change processes and the development of sustainable innovations can support mitigating market risks.

Moreover, in all three cases consumers are a critical change element. Beckenbach and Nill (2005, p. 79) were able to show in their study that products (and to a lesser extent, processes) are 'an effective contribution to the destabilisation of the ecological problematic lock-ins, at least concerning the catalyst of searching processes and the discovery of new techno-economical markers', the heterogeneous demands and related 'eco-niches'. Many consumers, however, maintain well-established consumption patterns and only recognise with difficulty the advantage of sustainable behaviour. As their behaviour considerably governs the market's demands new instruments of knowledge transfer and behaviour control are required here. The companies can achieve important contributions with proactive marketing and enlightenment campaigns.

External factors are often important in starting and catalysing sustainable change processes and for reinforcing sustainability-related actions. Hence, state regulation should focus on specific sustainability issues or fields by providing incentives. However, policy-makers should consider regulatory as well as market-based instruments and strengthen the negotiated (environmental) agreements, such as the provision of incentives or the establishment of multi-stakeholder groups. New institutions and methods to develop the consumers' awareness of environmental and sustainability-related issues should be implemented by the government. In addition, new strategies and means are required to enforce new activities and initiate changes in consumption decisions. Therefore, extensive social and cultural change is necessary, and a process that is likely to take some time.

References

Alvesson, M. (2005). *Understanding organizational culture*. London: Sage.

Aragón-Correa J. A., & Matías-Reche, F. (2005). Small firms and natural environment: A resource-based view of the importance, antecedents, implications and future challenges of the relationship. In S. Sharma & J. A. Aragón-Correa (Eds.), *Corporate environmental strategy and competitive advantage* (pp. 96–114). Cheltenham: Edward Elgar.

Argyris, C., & Schön D. A. (1996). *Organizational learning II, Theory, method and practice*. Reading: Prentice Hall.

Arnold, M. (2007). *Strategiewechsel für eine nachhaltige Entwicklung: Prozesse, Einflussfaktoren und Praxisbeispiele*. Marburg: Metropolis-Verlag.

Becke, G. (2003). Organisationales und ökologisches Lernen in kleinbetrieblichen Figurationen. In H. Brentel, H. Klemisch, Rohn. Lernendes Unternehmen (Hrsg.), *Konzepte und Instrumente für eine zukunftsfähige Unternehmens- und Organisationsentwicklung* (S. 193–215). Opladen: Leske + Budrich.

Beckenbach, F., & Nill, J. (2005). Ökologische Innovationen aus der Sicht der evolutorischen Ökonomik. In Jahrbuch Ökologische Ökonomik, Band 4, *Innovationen und Nachhaltigkeit* (S. 63–86). Marburg: Metropolis-Verlag.

Bonsen, M. (2000). Führen mit Vision. *Der Weg zum ganzheitlichen Management*. Niederhausen: Falken.

Brockhoff, K. (1999). Strategieidentifikation und Strategiewechsel. In G. Wagner (Hrsg.), *Unternehmensführung, Ethik, Umwelt* (S. 210–225). Wiesbaden: Gabler.

Brockhoff, K., & Leker J. (1998). Zur Identifikation von Unternehmensstrategien. Zeitschrift für Betriebswirtschaft, *68*(11), 1201–1223. Wiesbaden: Gabler.

Burmann, C. (2001). Strategische Flexibilität und Strategiewechsel in turbulenten Märkten. *DBW – Die Betriebswirtschaft, 61*, 169–188. Stuttgart: Schäffer Poeschel.

Burmann, C. (2002). Strategische Flexibilität und Strategiewechsel als Determinanten des Unternehmenswertes. Wiesbaden: Gabler.

Cleff, T., & Rennings, K. (1999). Empirische Evidenz zu Besonderheiten und Determinanten von Umweltinnovationen. In K. Rennings (Hrsg.), *Innovationen durch Umweltpolitik* (S. 47–99). Baden-Baden: Namos.

Dutton, J. E., & Duncan R. B. (1987). The creation of momentum for change through the process of strategic issue diagnosis. *Strategic Management Journal, 8*, 279–295.

Dyllick, T. (1989). *Management der Umweltbeziehungen: öffentliche Auseinandersetzungen als Herausforderung*. Wiesbaden: Gabler.

Eisenhardt, K. M. (1989). Making fast strategic decision in high velocity environments. *Academy of Management Journal, 32*, 542–576.

Evans, J. S. (1991). Strategic Flexibility for High-tech Manoeuvres – A Conceptual Framework. In: Journal of Management Studies, 28/1, (S. 69–89). Oxford: Blackwell.

Fichter, K., & Arnold, M. (2005). Entstehungspfade und Strategietypen bei Nachhaltigkeitsinnovationen. In K. Fichter, N. Paech, & R. Pfriem (Hrsg.), *Nachhaltige Zukunftsmärkte – Orientierungen für unternehmerische Innovationsprozesse im 21. Jahrhundert* (S. 97–108). Marburg: Metropolis-Verlag.

Galavan, R. (2008). Strategy, innovation, and change: Challenges for management. Oxford: Oxford University Press.

Gassmann, O. (1999). Praxisnähe mit Fallstudienforschung. *Wissenschaftsmanagement, 3,* 11–16.

Hammer, R. M. (1998). *Unternehmensplanung. Lehrbuch der Planung und strategischen Unternehmensführung* (7. Aufl.). München: Oldenbourg R. Verlag GmbH.

Hart, S. L. (1995). A natural-resource-based view of the firm. *Academy of Management Review, 20*(4), 986–1014.

Hedberg, B. (1981). How organizations learn and unlearn?. In P. C. Nystrom & W. H. Starbuck (Hrsg.), *Handbook of organizational design* (S. 8–27). London: Oxford University Press.

Hill, C. W. L., & Jones, G. R. (2008). *Strategic management theory: An integrated approach* (8th ed.). Boston, MA: Houghton Mifflin.

Howaldt, J., Klatt, R., & Kopp, R. (2004). *Neuorientierung des Wissensmanagements: Paradoxien und Dysfunktionalitäten im Umgang mit der Ressource Wissen.* Wiesbaden: Gabler.

Klimecki, R. G. (1997). Führung in der Lernenden Organisation. *Unternehmensethik, Managementverantwortung und Weiterbildung* (Vol. VI, pp. 82–105). Neuwied: Luchterhand.

Lambrechts, O. (2008). Proactive and reactive strategies for resource-constrained project scheduling with uncertain resource availabilities. In: Journal of scheduling, Jg. 11, H. 2, (S. 121–136). Norwell, Mass.: Springer Science + Business Media.

Lenker, J. (2000). Die Neuausrichtung der Unternehmensstrategie. Tübingen: Mohr Siebeck.

March, J. G. (1991). Exploration and exploitation in organizational learning. *Organization Science, 2,* 71–87.

Mayring, P. (2002). *Einführung in die qualitative Sozialforschung* (5. Aufl.). Weinheim: Beltz.

Meyer, A. D., Goes, J. B., & Brooks, G. R. (1995). Organizations reacting to hyperturbulence. In G. P. Huber (Ed.), Glick. *Organizational change and redesign, ideas and insights for improving performance* (pp. 66–111). New York: Oxford University Press.

Minder, S. (2001). *Wissensmanagement in KMU: Beitrag zur Ideengenerierung im Innovationsprozess.* St. Gallen: KMU Verlag HSG.

Nadler, D. A. (1988). Organizational frame bending: Types of change in the complex organization. In R. H. Kilman & T. J. Covin (Eds.), *Corporate transformation* (pp. 64–83). San Francisco: Jossey-Bass.

Nagel, R., & Wimmer, R. (2002). Systemische Strategieentwicklung. *Modelle und Instrumente für Berater und Entscheider.* Stuttgart: Schäffer Poeschel.

Pfriem, R. (2005). Strukturwandel und die Generierung von Zukunftsmärkten als neue Wettbewerbsebene. In K. Fichter, N. Paech, & R. Pfriem (Hrsg.), *Nachhaltige Zukunftsmärkte – Orientierungen für unternehmerische Innovationsprozesse im 21. Jahrhundert* (S. 27–56). Marburg: Metropolis-Verlag.

Pfriem, R., & Schwarzer, C. (2004). Innenwelt und Außenwelt. Organisationsentwicklung und ökologische Unternehmenspolitik. In R. Pfriem (Hrsg.), *Unternehmen, Nachhaltigkeit, Kultur. Von einem, der nicht auszog, Betriebswirt zu werden* (S. 124–150). Marburg: Metropolis-Verlag.

Porter, M. E., & van der Linde, C. (1995). Toward a new conception of the environment-competitiveness relationship. *Journal of Economic Perspectives, 9*(4), 97–118.

Rajagopalan, N., & Spreitzer, G. M. (1996). Toward a theory of strategic change: A multilens perspective and integrative framework. *AMR, 22,* 48–79.

Reinhardt, F. L. (1998). *Environmental product differentiation: Implications for corporate strategy. California Management Review, 40*(4), 43–73.

Rohracher, H. (1999). Zukunftsfähige Technikgestaltung als soziale Innovation. In D. Sauer & C. Lang (Hrsg.). *Paradoxien der Innovation: Perspektiven sozialwissenschaftlicher Innovationsforschung* (S. 175–189). Frankfurt/New York: Campus.

Sauer, D. & Lang, C. (1999). *Paradoxien der Innovation.* Frankfurt/New York: Campus.

Schein, E. H. (2003). *DEC is dead, long live DEC: The lasting legacy of digital equipment corporation*. San Francisco, CA: Berrett-Koehler.

Schein, E. H. (2008). Creating and managing a learning culture: The essence of leadership. *Business leadership* (pp. 362–369). San Francisco, CA: Jossey-Bass.

Schmidt, S. J. (2005). *Unternehmenskultur: die Grundlage für den wirtschaftlichen Erfolg von Unternehmen* (2. Aufl.). Weilerswist: Velbrück.

Senge, P. M. (2002). *The fifth discipline fieldbook: Strategies and tools for building a learning organization*. London: Brealey.

Shankar, V. (2009). Proactive and reactive product line strategies: asymmetries between market leaders and followers. In: Marketing strategy; Vol. 3: Marketing-mix strategies – product strategy and promotion strategy, (S. 3–30). Los Angeles: Sage.

Sharma, S., & Vredenburg, H. (1998). Proactive corporate environmental strategy and the development of competitively valuable organizational capabilities. *Strategic Management Journal, 19*, 729–753.

Siebenhüner, B., & Müller, M. (2003). Mit Umweltpolitik zu nachhaltigen Lernprozessen. *Zeitschrift für Umweltpolitik und Umweltrecht* (S. 309–332).

Teece, D. J., Pisano, G., & Shuen, A. (1997). Dynamic capabilities and strategic management. *Strategic Management Journal, 18*, 509–533.

Walgenbach, P. (2000). *Die normgerechte Organisation*. Stuttgart: Schäffer Poeschel.

Yin, R. (1994). *Case study research, design and methods*. Thousand Oaks, CA.

Zajac, E. J., et al. (2000). Modeling the dynamics of strategic fit: A normative approach to strategic change. *Strategic Management Journal, 21*, 429–453.

Chapter 13
Strategic Alliances for Environmental Protection

Haiying Lin and Nicole Darnall

Abstract Existing scholarship regarding strategic alliances has been limited by the tendency to view alliance formation through a single theoretical lens and to focus solely on the economic aspects (e.g., acquisition of capabilities) of narrowly defined relationships. As yet, there has been little attention paid toward examining how strategic alliances—of all sorts—can address social, economic and environmental issues. This chapter addresses these concerns by integrating the resource-based view of the firm with institutional theory to assess firms' decisions to participate in a strategic alliance. Drawing on these motivations, this chapter articulates a framework to characterize strategic alliances based on their competency- and legitimacy-orientation. A conceptual model is then constructed to examine the extent to which these strategic alliances are likely to encourage firms to adopt more (or less) proactive environmental strategies.

Keywords Strategic alliances · Alliance formation · Environmental performance · Resource-based view · Institutional theory

13.1 Introduction

In the last decade, the increasing uncertainty and complexity of the global business environment have led to the rapid proliferation of strategic alliances. Between 2000 and 2002 alone, over 20,000 strategic alliances were formed worldwide (Martin, 2002). Related to the natural environment, corporations increasingly are using strategic alliances to address complex environmental issues like climate change because of the scale and uncertainty embedded in these issues.

H. Lin (✉)
University of Waterloo, Faculty of Environment, Business and Environment, EV1-231,
200 University Avenue West, Waterloo, Ontario, N2L 3G1 Canada
e-mail: h45lin@uwaterloo.ca

J. Sarkis (eds.), *Facilitating Sustainable Innovation through Collaboration*,
DOI 10.1007/978-90-481-3159-4_13, © Springer Science+Business Media B.V. 2010

While previous research has recognized the importance of strategic alliances, these studies have had a strong tradition of assessing the economic aspects of inter-firm relationships (e.g., Mitchell & Singh, 1996). However, strategic alliances also involve cross-sector partnerships, and alliances of all sorts have been formed not only to address economic concerns, but also complex environmental issues. Additionally, previous scholarship has tended to treat strategic alliances as a dichotomous variable with participation relative to non-participation, thus failing to appreciate important nuances about their formation. For instance, some alliances may develop because of external institutional pressures, whereas others may form because of new market opportunities. These variations may lead to significant differences in an alliance's ability to accomplish meaningful environmental improvements.

Understanding these issues is important for policy-makers and non-government organizations alike because these entities are increasingly relying on strategic alliances as self-regulation mechanisms for firms to proactively manage their environmental problems. By recognizing which types of strategic alliances lead to more meaningful environmental outcomes, these organizations may be in a better position to shift their resources accordingly.

This chapter poses two research questions: (1) What types of strategic alliances are formed? (2) Which of these strategic alliances encourage firms to adopt more proactive environmental strategies? To attend to the first question, this paper integrates the resource-based view of the firm (RBV) with institutional theory to assess the variations in firms' motivations to participate in a strategic alliance. The second question is addressed by constructing a conceptual model that assesses how different types of strategic alliances are more likely to address complex environmental issues.

13.2 Understanding Strategic Alliance Formation

Strategic alliances are short- or long-term voluntary collaborations between organizations involving exchange, sharing or co-development of products, technologies and services to pursue a common set of goals or to meet a critical business need (Dacin, Oliver, & Roy, 2007; Gulati, 1998). In spite of their emphasis on collaboration, organizations that form strategic alliances retain their initial identities.

Strategic alliances can be inter-firm and cross-sector alliances. Inter-firm business alliances are established among two or more firms. For instance, since 2003 BP has partnered with DuPont to develop, produce and market the next generation of biofuels. By contrast, cross-sector alliances are partnerships between two or more organizations with fundamentally different governance structures and missions (Rondinelli & London, 2003). For instance, the U.S. Climate Action Partnership (USCAP) is a cross-sector alliance that was formed by 10 U.S.-based firms and four environmental non-government organizations (NGOs). This alliance seeks to establish a mandatory U.S. cap-and-trade program for carbon dioxide emissions.

To understand the reasons for why all sorts of strategic alliances are formed, we draw on literature on the RBV and institutional theory.

13.2.1 Resource-Based Explanations

RBV emphasizes the importance of firms' internal resources and competencies in explaining firm heterogeneity and competitive advantage (Prahalad & Hamel, 1990). Within the strategic alliance context, competitive advantages are derived from access to idiosyncratic resources, especially tacit knowledge-related resources from other organizations within an alliance setting (Das & Teng, 2000). When idiosyncratic resources/competencies are absent in the market, strategic alliances can help to combine complementary assets owned by different organizations (Hagedoorn, 1993) to develop valuable organizational competencies.

Given the ambiguities and uncertainty associated with environmental issues, strategic alliances can facilitate the flow of valuable information and opportunities to participating firms. By promoting organizational learning, firms may increase their ability to recognize and evaluate technological innovations in the marketplace. Improved organizational learning can mobilize firms to develop, acquire, and utilize their knowledge-based capabilities in a more effective way. Doing so helps firms develop different interpretations of new and existing information under conditions of ambiguity and uncertain information (Sharma & Vredenburg, 1998). Since the value of firm-level resources tends to dissipate over time as competitors replicate successful strategies, higher-order learning among participants can build up a capability for continuous environmental innovation, which can lead to sustained competitive advantage. (Sharma & Vredenburg, 1998)

Another way in which strategic alliances may create competitive advantage opportunities is that they offer a vehicle for some firms to shift existing practices toward creating the next-generation (Hamel, 1991) of technologies and business models. Radical repositioning of this sort is referred to as "creative destruction" and involves substituting existing unsustainable technologies with radically improved technologies that are environmentally friendly (Kemp, 1994). In creating next-generation technologies and business models, strategic alliances bring together unconventional partners and stakeholders, and examine emerging technologies and trends in product markets, with an eye toward creating new alternatives to existing products.

Finally, some strategic alliances foster societal opportunities (Eisenhardt & Schoonhoven, 1996). These alliances involve a team of like-minded individuals and organizations (Larson, 2000) who view achieving social and economic goals as being compatible and best achieved through mutual collaboration. For instance, firms that join these alliances may establish industry social codes of conduct that require industry participants to address their social impacts in a more robust way.

Such actions can place competitors at a competitive disadvantage while benefiting the environment (Etzion, 2007; Reinhardt, 1998). Other firms may align with regulators to improve their environmental performance. Along the way, these firms may foster good will with regulators and increase trust to such a degree that they can influence the environmental policy agenda (Darnall et al., 2008a). Related to climate change, industry leaders may align to support more stringent greenhouse gas mandates and force their competitors to follow suit. For instance, the Pew Center on Global Climate Change involved seven chemical and energy intensive companies to lobby the government for "early crediting" of firms' voluntary reductions of carbon dioxide and other greenhouse gases. Strategic alliances such as these greatly enhance firms' abilities to confer "above-normal" competitive advantages

Regardless of participants' motives, the specialized skills and competencies that result from resource-based motivations are anticipated to yield strategic alliances that are decentralized, firm-specific, knowledge-based and socially complex. We therefore term strategic alliances that are borne out of resource-based motivations to be *competency-oriented alliances*.

13.2.2 Institutional Explanations

While RBV provides one explanation regarding firms' motivations to participate in strategic alliances, institutional factors may also have an important role. Institutional theory posits that rules, norms, and values exert pressures on firms within a common setting to adopt similar practices and structures (DiMaggio & Powell, 1983) in an effort to gain social legitimacy and enhanced survival prospects (Meyer & Rowan, 1977). Decisions to participate in a strategic alliance are shaped by these institutional pressures. Such pressures arise from regulators, markets and society (Hoffman, 2000).

Regulatory pressures involve coercive legal mandates for organizations to adhere to regulations, rules and norms (Oliver, 1991). Firms that fail to adhere to these pressures risk obtaining non-compliance penalties, revocation of permit approvals, and unwanted media attention. In an attempt to seek approval from regulatory stakeholders, firms may strategically align with regulators for whom they depend for legal, physical, financial or reputation capital (Baum & Oliver, 1991; Dacin et al., 2007). The fact that firms' strategically align with regulators demonstrates that businesses are both influenced by government policies and actively involved in shaping their contexts and contesting, remaking, and redefining their institutional constituencies (Oliver, 1991).

Other types of institutional pressures arise from industry constituents. Perceived environmental uncertainty and social pressures (Baum & Oliver, 1991) encourage firms within the same industry to collaborate on specific environmental issues. For instance, in order to hedge their risk of an upcoming climate change policy, some industry participants may be motivated to align to manage their external constituencies collectively and work closely with political groups in an effort to influence public policy in a way that benefits existing products and processes. Moreover, to enhance industry-wide legitimacy, representative firms within an industry may align to explore technologies and solutions to ensure the legitimate operation of their current practices. These alliances share risk and investment among partners (Ring & Van de Ven, 1992) through economies of scale and scope. For instance, in response to institutional pressures related to climate change, the world's 10 largest coal and energy companies came together to explore the applicability of clean coal technology. Their alliance pooled together $1 billion to design, build and operate the world's first coal-fueled, near-zero emissions power plant.

Community constituents also exert institutional pressures on firms that may influence their decision to participate in a strategic alliance. As public concerns about environmental degradation rise, community constituents (especially environmental NGOs) are playing increasingly important roles. These individuals and groups can mobilize public sentiment, alter accepted norms, shift firms' environmental perceptions and imposing new roles to the firms (Hoffman, 2000), especially when they manage to align with influential constituents, such as regulators and investors to advance their agenda. For instance, in 1997 Environmental Defense Fund (EDF) published a report entitled *Toxic Ignorance*, which identified a lack of publicly available data on the chemicals produced in the highest production volumes. This report and the public attention it created put significant pressure on the chemical industry to respond. The pressure motivated the industry's trade association, American Chemistry Council, to partner with Environmental Protection Agency (EPA) and EDF to initiate the cross-sector alliance, High Production Volume Chemical Challenge Program. This alliance encouraged chemical firms to collect, summarize, compile and evaluate their existing chemical data, in addition to undertaking additional testing if necessary (Kent, 2004).

Whether regulatory, industry, or community-based, institutional pressures encourage firms to participate in strategic alliances to maintain or increase their social legitimacy. Yielding to these pressures can improve partnering firms' images, reputations, resources and market access, which in turn may enhance firms' chance of survival and improve their strategic market position (Dacin et al., 2007). We therefore term strategic alliances that are borne out of institutional pressures to be *legitimacy-oriented alliances*.

13.3 Dynamic Stategic Alliance Orientation

While individually RBV and institutional theory lend knowledge about the reasons why strategic alliances form, the integration of these two perspectives sheds more light on how these alliances are configured. We anticipate that firms participate in strategic alliances that are competency- or legitimacy-oriented, because of the particular societal or business issue confronting them. However, these societal and business issues may shift over time. Firms consequently respond in a dynamic way based on this change in context.

For instance, firms that participate in competency-oriented alliances develop innovations to enhance future business opportunities. These alliances can help speculating firms to obtain greater legitimacy with regulators and other social constituents. The outcome may be the establishment of tighter regulations in participants' favor or new industry standards that competing firms must adhere to, else risk lose their competitiveness. In response, competing firms may align and form legitimacy-oriented alliances to imitate the practices of industry first-movers. This situation exemplifies how strategic alliances can be seen as "experiments in institution building" (Osborn & Hagedoorn, 1997) that explain why common alliance practices emerge, are copied over time, and eventually become generally accepted practice.

Additionally, legitimacy-oriented alliances can help to facilitate information sharing and best-practices imitation among the participants. Firms therefore demonstrate various extents of incremental improvements and that can expand their knowledge capacities to a greater degree, and better position them to participate in competency-oriented alliances at a later time.

The case of BP exemplifies the dynamic orientation of firms' participation in strategic alliances. In the early 1990s, BP questioned the business opportunities related to climate change. As a consequence, it chose to join with other oil companies to form the Global Climate Coalition (GCC). This industry-based legitimacy-oriented alliance opposed ratification of the Kyoto Protocol by seeking to justify the industry's existing fossil fuel-based business practices. BP changed its position, however, in the late 1990s when technical improvements emerged and public concern about climate change was increasing. Because the company began to identify business opportunities associated with being proactive in its climate change position, the company shifted its alliance orientation to participate in competency-oriented alliances. BP became the first company to leave the GCC when it aligned with the University of California, in addition to DuPont, GE, and other firms to explore alternative energy solutions. Additionally, BP aligned with nine leading firms and NGOs to advocate that policy makers institute early carbon crediting and stringent mandates on allowable carbon thresholds. BP's aim was to disadvantage its competitors by forcing new industry standards. BP's efforts have helped increase regulatory pressures on ExxonMobil and other competitors by forcing them to soften their defiant stance toward climate change and consider how alternative energy technologies can help address the problem.

This example illustrates that participation in strategic alliances is not a binary choice in that companies either participate in either competency- or legitimacy-focused alliances. Rather, a company can participate in both types of strategic alliances at different points in time, depending on the business and societal issues confronting them. The example also illustrates that a firm's participation in strategic alliances may have consequences for other firms within that company's network or industry.

13.4 Relationship Between Strategic Alliances and Firms' Adoption of Environmental Strategies

The previous section discusses how institutional pressures and resource-based factors relate to the formation of different types of strategic alliances. This section extends the discussion by explaining how different types of strategic alliances facilitate the adoption of various environmental strategies.

13.4.1 Types of Environmental Strategy

Many scholars (e.g., Hart, 1995; Roome, 1992) have developed typologies of corporate environmental strategy. A commonality among them is that they recognize that corporate postures toward the natural environment range from reactive to proactive (e.g., Aragon-Correa, 1998).

A reactive posture is a response to changes in environmental regulations and stakeholder pressures that involve defensive lobbying and investments in end-of-pipe pollution control technologies (Aragon-Correa & Sharma, 2003, p. 73). Such technologies focus on addressing pollution after it has been created rather than eliminating waste before it is produced (Jones & Klassen, 2001). Related to climate change, carbon sequestration involves the separation of carbon dioxide from industrial and energy-related sources. Carbon dioxide is then transported to a storage location and isolated from the atmosphere (International Panel on Climate Change, 2005). This practice is considered reactive since it involves capturing carbon dioxide after it has produced. Similarly, the practice of converting waste to electricity is an example of a reactive posture in that it uses waste after it has been produced.

By contrast, proactive postures involve anticipating future regulations and social trends by designing or altering operations, processes, and products to prevent (rather than merely ameliorate) negative environmental impacts (Aragon-Correa & Sharma, 2003). There are at least three types of proactive environmental practices—pollution prevention, product stewardship and clean technology—that comprise a company's proactive posture (Hart, 1995).

Pollution prevention reduces pollution generation at the source before it is produced through better housekeeping, material substitution, recycling, or process innovation (Hart, 1995). Compared to end-of-pipe pollution control, pollution prevention focuses on extracting and using natural resources more efficiently, generating products with fewer harmful components, minimizing pollutant releases to air, and water and soil during manufacturing and product use (Organisation for Economic Co-operation and Development, 1995). In the area of climate change, one example of a pollution prevention practice is decarbonization from coal to gas. This process reduces the amount of carbon emitted per unit of primary energy by substituting natural gas for coal, which reduces carbon emissions per unit of electricity by half (Anderson & Newell, 2004). Similarly, cogeneration (combined heat and power) is an energy efficient technology that combines the usage of a power station to simultaneously generate both electricity and useful heat. Improved fuel economy, improved power plant efficiency and more efficient buildings are the three other pollution prevention options that address climate change concerns (Pacala & Socolow, 2004).

Like pollution prevention, product stewardship focuses on improving a firm's existing products. However, it extends the firm's reach by looking beyond organizational boundaries to individuals and organizations who are involved in a product's life cycle. Firms that undertake product stewardship assess the environmental performance of their products from raw material access, through production processes, to product use and disposal of used products (Hart & Milstein, 2003). Related to climate change, product stewardship options involve smart design/life cycle management for energy savings. They also include green supply chain management practices, which involve firms collectively considering the environmental attributes of their suppliers to avoid unnecessary environmental risks (Klassen & Whybark, 1999). By asking that their suppliers continually improve their environmental performance, firms can reduce the risk of inheriting environmental problems and minimize potential long-term environmental liabilities associated with their product inputs (Darnall et al., 2008b).

Clean technology refers not to the incremental product and process improvements associated with pollution prevention, but to innovations that leapfrog standard routines and knowledge (Hart & Milstein, 2003). Companies that pursue these efforts engage external stakeholders and build partnerships with nontraditional stakeholders such as environmental groups, consumer groups, and other companies, to acquire new competencies, knowledge, and vision (London, Rondinelli, & O'Neill, 2005). It includes such disruptive technologies as genomics, biomimicry, information technology, nanotechnology and renewable energy applications that enable firms to shift away from traditional fossil fuel economies. Other examples include renewable energy applications such as biomass, solar thermal and photovoltaic, wind, hydropower, ocean thermal, geothermal and tidal power generation (Johansson, Kelly, Reddy, & Williams, 1993).

The above discussion suggests that proactive environmental practices are not necessarily independent of one another in that implementing a product stewardship program often requires firms to have a strong understanding of pollution prevention or risk significantly greater implementation costs (Darnall & Edwards, 2006). Additionally, there is a path dependence for firms that wish to implement clean technology such that they may need skills in pollution prevention and product stewardship to innovate for the environment in a meaningful way (see Fig. 13.1). Moving along the proactive environmental strategy path requires greater resource accumulations and reconfigurations. Such movement also requires that firms consider changing their business models, technologies, operation processes, and performance objectives (Sharma & Henriques, 2005), in addition to significant investments in knowledge-based organizational systems and practices. As such, scholars (e.g., Ashford, 1993) emphasize the potential conceptual, technical and organizational

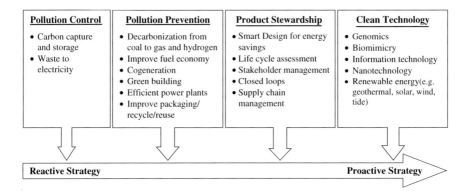

Fig. 13.1 Corporate environmental strategies that mitigate climate change

barriers that prevent firms from becoming more environmentally proactive. Strategic alliances may be one mechanism to overcome these obstacles.

13.4.2 Strategic Alliances and Environmental Strategy

Proactive environmental strategies often require firms to commit resources toward initiating significant changes in processes or new production technologies (Hart & Ahuja, 1996). They necessitate a long-term vision shared among all relevant stakeholders, significant employee involvement, cross-disciplinary coordination and integration, a strong moral leadership and forward-thinking managerial style (Shrivastava, 1995). Our belief is that firms which participate in competency-oriented alliances are positioned to adopt environmental strategies that are more proactive than firms that participate in legitimacy-oriented alliances.

Competency-oriented alliances help participants to improve their internal learning by combining complementary competencies from heterogeneous partners. These efforts can trigger environmental innovation. Compared to the legitimacy-oriented alliances that target incremental process innovation, the innovation spurred by competency-oriented alliances tends to create far-reaching, radical and transformative changes to business models and markets. Such change includes the promotion of new products, formulation of new markets and identification of new means of sustainably servicing existing markets (Etzion, 2007).

Since the economic returns associated with proactive environmental strategies may not be directly visible or may occur only in the long term, firms are discouraged from acquiring knowledge and shifting managerial attitudes toward implementing proactive environmental strategies (Ashford, 1993). However, the higher-order learning that is developed by engaging in competency-oriented alliances can help corporate managers acquire knowledge of these long term benefits, inform attitudes, and subsequently build an internal commitment toward adopting more proactive environmental strategies. One way in which firms participating in competency-oriented alliances acquire this knowledge and shift managerial perceptions of environmental problems is by involving heterogeneous partners. Heterogeneous partners, which may include nonprofit social organizations and environmental NGOs, can provide stronger complementary assets for innovation or entry of new markets than homogeneous partners. Such diversity is also important for creating the innovation and new market entry that are a focus of competency-oriented alliances.

For instance, BP America, DuPont, Alcoa, and other firms have aligned with World Resource Institute (WRI), a Washington, DC-based NGO, in an effort to deploy climate-friendly technologies, market green power, and campaign for early-crediting of greenhouse gas reductions. These alliances have included the U.S. Climate Action Partnership, The Green Alliances, and the Greenhouse Gas Protocol Initiative. Additionally, General Electric, Johnson & Johnson, and 11 other firms have aligned with WRI to initiate the Climate Northeast Partnership in an effort to develop strategies for business to thrive in a carbon-constrained economy. The primary goal of each of these alliances was to expand renewable energy technologies and increase corporate demand and markets for renewable energy. These examples illustrate that combining complementary competencies from heterogeneous partners tend to create greater opportunities for transformative changes to business models and markets that stem from more proactive environmental strategies. By partnering with a broad array of stakeholders, committing resources toward significant changes in processes or new production technologies, and having leaders to see these efforts through, competency-oriented alliances are poised to improve the environment in a meaningful way.

13.4.2.1 Proposition 1: Competency-Oriented Alliances Tend to Associate with More Proactive Environmental Strategies

Compared to competency-oriented alliances, the primary driver in the formation of legitimacy-oriented alliances is not competency-building but achieving external credibility. Firms that participate in legitimacy-oriented alliances meet (but not exceed) social expectations. As a result, firms participating in legitimacy-oriented alliances firms may reproduce, imitate and sustain legitimate organizational structures, activities and routines and become resistant to change over time. Firms that participate in these alliances also tend to be skeptical of new technology until sufficient experience has developed within its industry. While this skepticism allows managers to reduce short-term risk, it also causes firms to miss profit opportunities related to the implementation of more proactive environmental strategies (King & Lenox, 2002).

Additionally, firms participating in legitimacy-oriented alliances are more likely to have employees who are motivated to justify entrenched organizational habits and routines. This rigidity creates a cycle that limits cross-functional cooperation that might introduce innovative environmental solutions. Combined, these factors create substantial impediments to adopting advanced innovations and more proactive environmental strategies (Ashford, 1993). For these reasons, we propose that legitimacy-oriented alliances are more likely to be associated with a less proactive environmental strategy.

13.4.2.2 Proposition 2: Legitimacy-Oriented Alliances Tend to Associate with Less Proactive Environmental Strategies

Figure 13.2 summarizes the relationships we have discussed. It illustrates how corporate motivations lead to firms choosing between participation in competency- and legitimacy-oriented alliances. Neither type of alliance is separate in that companies can participate in competency- or legitimacy-focused depending on the social and business context. Figure 13.2 further illustrates that alliance choice is associated with a continuum of environmental strategy outcomes that range from reactive to proactive.

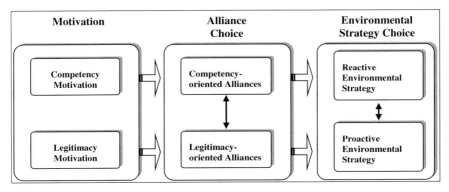

Fig. 13.2 Relationship between firms' motivations to participate in a strategic alliance and subsequent choice of environmental strategy

13.5 Conclusion

This chapter contributes to our understanding of strategic alliances by developing a framework to assess alliance formation. This framework goes beyond treating strategic alliances as a dichotomous variable and appreciates the important nuances associated with their formation. We suggest that strategic alliances are formed because of firms' motivations to enhance their resources and capabilities, in addition to their desire to address institutional pressures. Variations in these motivations lead to a continuum of strategic alliances with competency-oriented alliances at one end and legitimacy-oriented alliances at the other. We posit that these variations lead to significant differences in each alliance's ability to accomplish meaningful environmental improvements. In particular, we suggest that competency-oriented alliances tend to associate with more proactive environmental strategies, whereas legitimacy-oriented alliances tend to associate with less proactive environmental strategies.

Future research would benefit by empirically examining whether or not firms that participate in competency-oriented alliances benefit the environment in a more meaningful way. Knowledge of this relationship has important implications for policy-makers and NGOs alike, since many of these individuals and groups are endorsing (and even developing) strategic alliances to advance their environmental protection goals. By appreciating which strategic alliances are more likely to lead to more meaningful environmental outcomes, policy makers and NGOs can shift their attention accordingly. For instance, since legitimacy-oriented alliances may be associated with less proactive environmental strategies, they may be less likely to lead to meaningful environmental outcomes than competency-oriented alliances. As such, policy makers and NGOs may achieve stronger environmental outcomes by not simply pressuring for environmental change among the regulated community, but also by aligning with businesses to foster stronger learning and innovation that leads to more ambitious environmental outcomes. Our hope is that the discussion presented in this chapter offers sufficient reason for future scholarship to consider these issues further.

References

Anderson, S., & Newell, N. (2004). Prospects for carbon capture and storage technologies. *Annual Review of Environmental Resources, 29*, 109–142.

Aragon-Correa, J. A. (1998). Strategic proactivity and firm approach to the natural environment. *Academy of Management Journal, 41*, 556–567.

Aragon-Correa, J. A., & Sharma, S. (2003). A contingent resource-based view of proactive corporate environmental strategy. *Academy of Management Review, 28*(1), 71–88.

Ashford, N. A. (1993). Understanding technological responses of industrial firms to environmental problems: Implications for government policy. In J. Schot & K. Fischer (Eds.), *Environmental strategies for industry: International perspectives on research needs and policy implications* (pp. 277–310). Washington, DC: Island Press.

Baum, J., & Oliver, C. (1991). Institutional linkages and organizational mortality. *Administrative Science Quarterly, 36*, 187–218.

Dacin, M. T., Oliver, C., & Roy, J. (2007). The legitimacy of strategic alliances: An institutional perspective. *Strategic Management Journal, 28*, 169–187.

Darnall, N., & Edwards, Jr., D. (2006). Predicting the cost of environmental management system adoption: The role of capabilities, resources and ownership structure. *Strategic Management Journal, 27*(2), 301–320.

Darnall, N., Henriques, I., & Sadorsky, H. (2008a). Do environmental management systems improve business performance in an international setting?. *Journal of International Management, 14*, 364–376.

Darnall, N., Jolley, G. J., & Handfield, R. (2008b). Environmental management systems and green supply chain management: Complements for sustainability?. *Business Strategy and the Environment, 17*(1), 30–45.

Das, T. K., & Teng, B. S. (2000). A resource-based theory of strategic alliances. *Journal of Management, 26*, 31.

DiMaggio, P. J., & Powell, W. W. (1983). The iron cage revisited: Institutional isomorphism and collective rationality in organizational fields. *American Sociological Review, 48*, 147–160.

Eisenhardt, K. M., & Schoonhoven, C. B. (1996). Resource-based view of strategic alliance formation: Strategic and social effects in entrepreneurial firms. *Organization Science, 7*, 136–150.

Etzion, D. (2007). Research on organizations and the natural environment, 1992–present: A review. *Journal of Management, 33*(4), 637–664.

Gulati, R. (1998). Alliances and networks. *Strategic Management Journal, 19*(4), 293–317.

Hagedoorn, J. (1993). Understanding the rationale of strategic technology partnering: Interorganizational modes of cooperation and sectoral differences. *Strategic Management Journal, 14*(5), 371–385.

Hamel, G. (1991). Competition for competence and inter-partner learning within international strategic alliances. *Strategic Management Journal, Summer Special Issue 12*, 83–103.

Hart, S. (1995). A natural-resource-based view of the firm. *Academy of Management Review, 20*(4), 986–1014.

Hart, S. L., & Ahuja, G. (1996). Does it pay to be green? An empirical examination of the relationship between emission reduction and firm performance. *Business Strategy and the Environment, 5*(1), 30–37.

Hart, S. L., & Milstein, M. B. (2003). Creating sustainable value. *Academy of Management Executive, 17*(2), 56–69.

Hoffman, A. (2000). *Competitive environmental strategy: A guide to changing the business landscape*. Washington, DC: Island Press.

International Panel on Climate Change. (2005). Carbon dioxide capture and storage, summary for policymakers, a special report of working group III of the Intergovernmental Panel on Climate Change. *IPCC Special Report*.

Johansson T. B., Kelly, H., Reddy, A. K. N., & Williams, R. H. (Eds.). (1993). *Renewable energy: Sources for fuels and electricity*. Washington, DC: Island Press.

Jones, N., & Klassen, R. D. (2001). Management of pollution prevention: Integrating environmental technologies in manufacturing. In J. Sarkis (Ed.), *Greener manufacturing and operations: From design to delivery to takeback* (pp. 56–68). Sheffield: Greenleaf Publishing.

Kemp, R. (1994). Technology and the transition to environmental sustainability: The problem of technological regime shifts. *Futures, 26*(10), 1023–1046.

Kent, D. J. (2004). Status and trends of the HPV program in the USA and Europe. *Regulatory Affairs Bulletin, 94*, 1–7.

Klassen, R. D., & Whybark, D. C. (1999). The impact of environmental technologies on manufacturing performance. *Academy of Management Journal, 42*, 599–615.

King, A., & Lenox, M. (2002). Exploring the locus of profitable pollution reduction. *Management Science, 48*(2), 289–299.

Larson, A. L. (2000). Sustainable innovation through an entrepreneurship lens. *Business Strategy and the Environment, 9*(5), 304–317.

London, T., Rondinelli, D. A., & O'Neill, H. (2005). Strange bedfellows: Alliances between corporations and nonprofits. In O. Shenkar & J. Reuer (Eds.), *Handbook of strategic alliances* (pp. 353–366). Thousand Oaks, CA: Sage Publications.

Martin, J. G. (2002). The growing use of strategic alliances in the energy industry. *Oil, Gas and Energy Law Journal, 27*(1), 28–57.

Meyer, J. W., & Rowan, B. (1977). Institutionalized organizations: Formal structure as myth and ceremony. *American Journal of Sociology, 83*, 340–363.

Mitchell, W., & Singh, K. (1996). Survival of businesses using collaborative relationships to commercialize complex goods. *Strategic Management Journal, 17*(3), 169–195.

Oliver, C. (1991). Strategic responses to institutional processes. *Academy of Management Review, 16*, 145–179.

Organisation for Economic Co-operation and Development. (1995). *Technologies for cleaner production and products*. Paris: OECD.

Osborn, R. N., & Hagedoorn, J. (1997). The institutionalization and evolutionary dynamics of interorganizational alliances and networks. *Academy of Management Journal, 40*, 261–278.

Pacala, S., & Socolow, R. (2004). Stabilization wedges: Solving the climate problem for the next 50 years with current technologies. *Science, 305*, 968–972.

Prahalad, C. K., & Hamel, G. (1990). The core competence of the corporation. *Harvard Business Review, 68*(3), 79–91.

Reinhardt, F. L. (1998). Environmental product differentiation: Implications for corporate strategy. *California Management Review, 40*(4), 43.

Ring, P. S., & Van de Ven, A. H. (1992). Structuring cooperative relationships between organizations. *Strategic Management Journal, 13*(7), 483–498.

Rondinelli, D. A., & London, T. (2003). How corporations and environmental groups cooperate: Assessing cross-sector alliances and collaborations. *Academy of Management Executives, 17*(1), 61–76.

Roome, N. (1992). Developing environmental management strategies. *Business Strategy and the Environment, 1*, 11–24.

Sharma, S., & Henriques, I. (2005). Stakeholder influences on sustainability practices in the Canadian forest products industry. *Strategic Management Journal, 26*(2), 159–180.

Sharma, S., & Vredenburg, H. (1998). Proactive corporate environmental strategy and the development of competitively valuable organizational capabilities. *Strategic Management Journal, 19*, 729–753.

Shrivastava, P. (1995). Environmental technologies and competitive advantage. *Strategic Management Journal, 18*, 183–200.

Chapter 14
Towards Sustainability Through Collaboration Between Industrial Sectors and Government: The Mexican Case

María Laura Franco-García and Hans Th. A. Bressers

Abstract This chapter discusses the extent to which Dutch experiences with nego-tiated agreements between firms and public authorities could be used as a tool to improve environmental policies and foster collaboration and innovation for sustain-ability in Mexico. The Mexican context is analysed both in terms of perceived effectiveness of environmental regulation/existing voluntary agreements and in terms of attitudes and opinions of key players in the Mexican Industry regarding feasibility of negotiated agreements. Our findings show that there is good recep-tivity to the use of negotiated agreements both from the point of view of policy makers and industry leaders. The comparison with Dutch experiences shows no important gap between Mexican business leaders' expectations regarding results in terms of efficiency gains and positive side effects and the results obtained by negoti-ated agreements in the Netherlands. Mexico benefits from a history of trust and fair play between the industrial sector and the government; homogeneity or clear lead-ership in polluting industrial sectors. Polluting firms are also concerned with their public Image and there is a widespread belief that the government will resort to other measures if negotiation fails. All the latter factors, which were determinant of success in The Netherlands, support the feasibility of using negotiated agreements as a collaborative strategy towards sustainability in Mexico.

Keywords Negotiated agreements · Mexico · Business-government collaboration · Polluting firms · Corporate voluntary initiatives

14.1 Introduction

The environmental performance of business in Mexico has significantly improved in the last decade thanks to the implementation of several regulatory instruments

M.L. Franco-García (✉)
Centre for Clean Technology and Environmental Policy (CSTM), University of Twente, Enschede, 7500 AE, The Netherlands
e-mail: m.l.francogarcia@utwente.nl

J. Sarkis (eds.), *Facilitating Sustainable Innovation through Collaboration*,
DOI 10.1007/978-90-481-3159-4_14, © Springer Science+Business Media B.V. 2010

and also due to some economical incentives for those companies certified as "clean industry" (mostly those considered of "large size"). It is also important to mention that large companies generally have already had a good quality environmental system implemented for a long time based on environmental and social goals that originate from their international corporation's head offices (OECD, 2003). From this point of view the enterprise's culture and financial situation seem to be key factors for environmental success because the large companies are more able to invest in environment control and pollution prevention systems. This was also reported by Medina-Ross in 2003 in the context of Corporate Voluntary Environmental Initiatives (VEI) in Mexico for the chemical sector.

A voluntary program such as "clean industry" is expected to have a "snowballing effect" through the industry and "cascade down" from large companies to SMEs. However, a significant number of small and medium size enterprises remain without "clean industry" certification, largely because they cannot comply with the legal requirements established by the environmental authorities (Tornel, 2007). Some of their problems include the perception of environmental regulation as very complex and difficult to understand; and the unclear distribution of responsibilities among the different governmental levels. As a consequence the enterprise does not directly know its rights and obligations to the local, state and federal governments in terms of environment regulation. Lack of trust and inadequate communication between authorities and firms are as well associated with this scenario of low adherence to voluntary approaches.

The Netherlands, on the other hand, is a country with excellent experiences in terms of environmental management and voluntary agreement implementation. The Netherlands has used negotiated agreements between authorities and business as a policy tool to improve trust and communication, thus having an important impact on the transparency of voluntary approaches, while fostering collaboration for innovation (Bressers & de Bruijn, 2005a).Thus, the use of negotiated agreements can be the key to improve increase the rate of industrial adherence voluntary policy environmental instruments in Mexico. However, it is arguable to what extent a policy-tool that had been developed for a specific European context can be transferred to the seemingly radically different Latin-American environment.

This chapter will use interviews and survey methods to evaluate the feasibility of carrying on negotiated agreements between business and authorities to develop voluntary approaches in the Mexican context. The group of businesses targeted in the analysis are Mexican industry leaders, influential companies with a strong position in the production chain and also with an active role in environmental policy negotiations. SME's are not well represented among them, but the industry leaders have an important influence on the SME's group behaviour through supply chain effects

In Section 14.2 we will briefly describe some aspects on the literature of voluntary and collaborative approaches – in particular negotiated agreements – that we perceive as relevant for the Mexican context. The latter is in turn analysed in Section 14.3. It will be shown that although direct regulation is strategically still the main instrument used in Mexican environmental policy, substantial use of collaborative and voluntary approaches has begun (Alvarez-Larrauri & Fogel, 2008). This could provide the basis for more ambitious agreements. Section 14.4 will describe the study's methodology while Section 14.5 (results) will present an assessment of the attitudes of a number of Mexican business leaders with environmental responsibilities regarding the use of negotiated agreements. Lastly, Section 14.6 will provide a short summary and conclusions.

14.2 Literature Review

Voluntary approaches are receiving increased attention, in and outside the European framework where they are most widely used. Recently two subsequent issues Policy Studies Journal devoted no less than 10 articles to this subject, in symposia edited by DeLeon and Rivera (2007, 2008). Berchicci and King (2007) debate the value of "self-regulatory institutions" for sustainability.

Among the policy instruments classified as voluntary, "negotiated agreements" have a special place, since they are what the name suggests: not entirely voluntary, but the result of real negotiations. The negotiations often concentrate on the share of responsibility a certain sector of industry will take in realising the countries' environmental objectives. In the Netherlands this strategy has been more often and systematically applied that anywhere else in the world (Glasbergen, 1998).

Negotiated agreements are defined as the "commitments undertaken by firms and sector associations, which are the result of negotiations with public authorities and/or explicitly recognised by the authorities" (EEA, 1997, p. 11). They can be regarded as a subspecies of "voluntary approaches" (Börkey & Lévêque, 2000). Unilateral commitments and public voluntary programs are two other forms of such approaches. Since the terminology of voluntary approaches is much more common, especially outside of Europe, it is important to realise that negotiated agreements are precisely what the name suggests: agreed upon commitments that stem from governmental and other pressures as well as industry's acceptance. This does not preclude that in some cases it is industry and not government who takes the initiative to commence talks. Industry will in these cases have recognised that in one way or another action will be required and prefers to be involved in the formulation of the program.

Notwithstanding its popularity, the direct environmental results of negotiated agreements generally prove to be mixed (see the numerous cases in Carraro & Lévêque, 1999; Croci, 2005; De Clercq, 2002; Delmas & Terlaak, 2001; Klok, 1989; Mol, Lauber, & Liefferink, 2000; Orts & Deketelaere, 2001; Rennings, Brockman, & Bergmann, 1997; Ten Brink, 2002). The benefits of the use of covenants are, therefore, not undisputed. Some studies are fairly critical about the effects voluntary approaches generate (for instance OECD, 2003). Combined approaches have also been evaluated in the US context where taxation and self regulation together with voluntary agreements are, according to authors, going to be extended (Lyon & Maxwell, 2003). The results in the Netherlands are however quite good. Very generally speaking these "mixed results" imply that for direct environmental results negotiated agreements are not necessarily better or worse than alternative instrumental strategies. They are better apt for some situations and worse for others (cf. Bressers & de Bruijn, 2005a) and their strength can be generally assessed as "a good way to attain the feasible environmental improvement", implying that it is less able to shift the boundaries of what is regarded as feasible. Their effectiveness proves to be highly dependent on their position in and linkages to the policy system as a whole (de Bruijn & Norberg-Bohm, 2005). So, while the results are quite good when used for sustainability in the Netherlands, which can be seen as the capital of negotiated agreements, it makes sense to carefully examine the "fit" in another context before just proposing to copy Dutch schemes.

What would be the reasons to make such a study worthwhile? We have derived them from the literature and mention them here. Later in Section 14.5 we will come back to them to discuss the same issues in the Mexican context based on a small survey.

First of all the situation regarding environmental policy in many countries, including Mexico, is not very favourable, especially when *implementation* is concerned. Negotiated agreements have the potential advantage that they often have an extended implementation structure that creates a permanent or at least regular platform for communication and integration. In this implementation structure there can be also room for a permanent advisory agency and for exchanges of experiences among the different sectors of industry. Such an implementation structure can create fruitful collaboration patterns that enhance implementation, in addition to the minimally required legal standards. In the Netherlands sector wide negotiated agreements are not replacing permitting and enforcement, but create a framework for them (apart from giving guidance for further developments; Bressers & de Bruijn, 2005b). In Fig. 14.1 this relationship is elaborated.

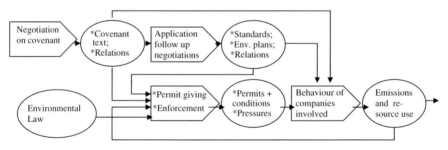

Fig. 14.1 Input-process-output model of environmental policy implementation combining individual regulation and sectoral negotiated agreements

A further potential advantage is that using negotiated agreements can not only achieve direct environmental benefits (with a certain level of *ambition*), but can also create flexibility that makes more *efficiency* possible and will tend to stimulate collective learning processes that build *positive side-effects* for future next steps. Examples mentioned in literature include: better relations, a more shared problem perception, and more knowledge exchange (De Clercq, 2002, pp. 44–45). The collaborative implementation style can in principle enable real innovations. In the Netherlands, respondents of the official evaluation study on the practice of negotiated agreements (Bressers, de Bruijn, & Lulofs, 2009) assessed that the *ambition* of negotiated agreements were indeed beyond existing regulation (75% agreed), beyond business as usual (86% agreed), implied real ecological innovation (68% agreed), and were guarded against free riders (72% agreed). In the Dutch evaluation study on the *efficiency* of negotiated agreements it was assessed that the negotiated agreements were minimising total costs (55% agreed), especially by creating improvements in phasing flexibility (75% agreed) and allocation of efforts (96% disagreed with the proposition that better allocation could have lowered costs). There was less support that they led to decreased administrative costs (48% agreed) and that new methods and technologies were developed (44% agreed). In fact the *positive side-effects* were the most impressive scores in the Dutch situation. We will compare them with Mexican business leader's expectations in Section 14.5.

Having said this, not only the potential effects count in the usability of the Dutch style negotiation agreements, but also their *feasibility* in a certain societal and political context. Previous studies in six European countries (De Clercq, 2002) later validated in the Netherlands (Bressers & de Bruijn, 2005a), showed four crucial factors that relate to the success and feasibility of negotiated environmental agreements. Table 14.1 shows the assessments (in %) for these factors in the Dutch evaluation study.

Table 14.1 Conditions for negotiated agreements' success in the Netherlands

In %	Entirely agree	Agree	Neutral	Disagree	Entirely disagree
Before the negotiation there was already trust between the sector and government	2	47	6	37	8
The representative sector organisation could negotiate on behalf of the member companies	31	39	2	26	2
The public image of the sector or its product is sensitive to environmental aspects	28	55		18	
The authorities saw it as a realistic option to use other instruments when negotiations would fail	16	47		37	

Thus, (1) potential to deal with failing implementation of command and control regulations, (2) own results, including ambition, efficiency and positive side-effects, and (3) the conditions for success and feasibility are the three main issues as seen from literature and empirical research. Before we discuss each of them in separate sections of Section 14.5 on the basis of opinions of business leaders in Mexico, we will first discuss the environmental policy context in Mexico and the existing role of voluntary approaches herein. An innovative approach never finds its setting without context and history. What is the basis on which a Mexican negotiated agreement approach could be grounded?

14.3 Environmental Management and "Voluntary" Approach in Mexico

Mexico has a quite complete regulatory framework in environmental themes for controlling industrial pollution (OECD, 1998) and a bureaucratic-administrative structure prepared for the design and implementation of environmental regulation (Mumme, 1998), which has been empowered since the Guadalajara explosion in April 1992. The creation of the "Procuraduría Federal de Protección Ambiental" (PROFEPA) is an example of that which stands for specifically controlling industrial pollution. PROFEPA established two strategies in order to implement the regulation: the "coercive" approach involving inspection and surveillance to assure the regulations' compliance; and the "voluntary" strategy which was labelled as "Programa Nacional de Auditoria Ambiental" (PNAA; PROFEPA, 2008).

The "coercive" approach faces some resource limitations (personnel and budget) to inspect and audit compliance because of the large number of potential industrial pollutant sources. The analysis between the number of auditing activities by year and the number of industries in Mexico reported by COPARMEX (Table 14.2) show that a small company will be audited with a probability of 1:700. Even while most resources are dedicated to audit large companies, it still would take 8 years to cover all of them. The monitoring and enforcement of environment regulation were previously indicated as "sporadic" by Dasgupta, Hettige, and Wheeler in 2000 in their research work about environmental improvement compliance in Mexico.

Table 14.2 Distribution of audits in the Mexican enterprise size groups (COPARMEX, 2005)

Denomination according to industry size	Number of employees	Distribution in enterprise size	(National Environmental Audit program) 2004 data
Micro	0–15	335500	483 (1:695)
Small	16–100	19500	169 (1:115)
Medium	101–250	3700	196 (1:19)
Large	> 250	2300	285 (1:8)
	Total number of enterprises	361000	1133

Enlarging the resources to increase governmental capacities will turn the process very costly, further more the command-control strategy by itself doesn't represent the most effective option, at present.

The national "voluntary" approach has provided a complementary strategy based on the implementation of EMS as ISO 14001. These voluntary international instruments have acceptance inside the industry, partially due to the economic pressures by international stakeholders (Jordan et al., 2003) and besides that, because companies move further towards acknowledging that there can be a win-win relationship and see the environmental performance as a competitive advantage (Sarkar, 2008;

Porter & van der Linde, 1995). On the other hand this might also result in the risk of environmental improvements only at the level of "low hanging fruits". Consequently the dilemma of using international instruments for the improvement of the national environmental situation remains, especially because there are few meaningful improvement indicators (Press, 2007).

In the particular case of the PNAA, it is important to mention that this program was originally an initiative from the business sector and was only supported by the Mexican government following this. The PNAA includes the Clean Industry certification, referred to in Spanish as "Industria Limpia". This distinction for the enterprise is seen as a reward that after scrutiny by a team of experts, it has demonstrated that the enterprise has invested in all necessary modifications, both technological and organisational, to comply with environmental regulations. In 2764 audits, Mexican companies invested US$2.155 billon in environment pollution control through the PNAA. This was reported by Alvarez-Larrauri and Fogel (2008) in their analysis of 10 years of experience in Mexico. Besides PNAA's success in promoting environmental control investments, successes were also reported in limiting environmental risk and motivating worker participation.

According to such data, voluntary agreements became a convenient option for enhancing environmental protection in countries like Mexico. Indeed, the PNAA has been successful in many aspects but the adherence rate has been growing quite smoothly, this is in particular for the small and medium size industries. Table 14.3 shows the comparison made by COPARMEX in terms of the number of "Clean Industry" and ISO 14000 certifications.

Table 14.3 Number of industries with Mexican and International certifications (both voluntary; COPARMEX, 2003)

Year	No. of industries with "Clean Industry" certification	No. of industries with "ISO 14000" certification
1997	11	886
1999	63	1345
2002	369	2212

Therefore, research in order to identify the implementation conditions of negotiated environmental agreements in the Mexican context is the aim of this work. Our reference was the Dutch Environmental Policy due to the excellent experiences in terms of environmental management and voluntary agreement implementation (see Table 14.1). Basically, our focus is centred on "how to increase the rate of industrial adherence to PNAA or some other voluntary policy environmental instruments". The methodology and results are described and analysed in the next chapters. Our general hypothesis is: using voluntary and negotiated instruments in the Mexican environmental policy will improve effective collaboration between government and industry sector.

14.4 Explorative Survey of Attitudes Towards the Application of Negotiated Agreements in Mexico

In the spring of 2008 a survey was sent by email to 60 Mexican business managers (level of plant manager or just below) that were believed to be responsible for environmental matters in their plants. The sample was selected by the Mexican national level Chamber of Commerce, based on which companies they had most contacts with on environmental policy matters. So the sample was deliberately selected to involve "opinion leaders" in this regard among Mexican businesses. When the response proved to be limited (in number, not unusually low in percentage), in some instances addresses were added on the indication of respondents already in the sample, so-called "snowballing". The questionnaire contains 11 questions, seven multiple choice questions (with some space to comment) and four blocks of in total 22 propositions to either agree or not on a five point scale. The response number was 16 in total. Later inquiries showed that a main cause for not answering was that in many cases the prospected respondents needed the permission of company headquarters to fill in external questionnaires, while additionally many busy business leaders categorically ignore questionnaires sent by email.

Nevertheless, the respondents represent a wide array of Mexican industries. They include the food industry, chemicals, metals, non-metal products, concrete, pharmaceutics, construction, glass and car parts sectors. Of the 16, 9 have a predominantly administrative function, like general management, 6 a predominantly technical function, like environmental management, and 1 a predominantly external relations function. Most are also quite senior in terms of number of years spent with the company. Four have worked 3 years or less with the company, four up to 6 years and eight more than 6 years. Most of them (13) are members of an environmental board or committee of their company. Six are members of an environmental committee of their sector of industry. Only one belongs to none of those. In addition two are members of government environmental committees and three are active members of NGOs.

The questions asked reflect the issues that were identified in the literature review in Section 14.2. They were about the evaluation of the present state of Mexican environmental policy and the options to improve the set of policy instruments or their *implementation*, the importance of several possible *ambitions, efficiency and side-effects* of negotiated agreements and about some conditions that impact on their *feasibility* and success. These questions were designed to be, as much as possible, identical to questions asked in the ex post evaluation study on the Dutch system of environmental negotiated agreement that was concluded a few years ago (Bressers & de Bruijn, 2005a, 2005b; Bressers et al., 2009; de Bruijn, 2003).

Because we could not be sure that all respondents had a clear as well as similar idea of the notion of negotiated agreements, we introduced the questionnaire with the following text:

"*Negotiated agreements* are defined as the 'commitments undertaken by firms and sector associations, which are the result of negotiations with public authorities and/or explicitly recognised by the authorities'. They can be regarded as a subspecies of 'voluntary approaches'. Unilateral commitments and public voluntary programs, like 'Industria Limpia', are other forms of such approaches. But compared to real voluntary approaches, it is much more oriented towards mid- and long-term improvements in environmental performance. It is not uncommon in Europe that industry itself takes the initiative to start negotiations. This makes sense, especially in cases when it seems inevitable that government will push one way or another for substantial environmental improvements. Negotiating and agreeing on a 5 or even 10 year schedule that fits normal business investment schemes might then be preferred over awaiting regularly changing top-down regulations. The advantage for government can be that environmental considerations start playing a role earlier in business' decision making processes and the advantage for business is that environmental requirements come less as a disruption of normal business processes. The negotiated agreement is often concluded at a higher scale level than individual companies, for instance a sector of industry in a certain state, and its progress followed by joint committees, in which mutual trust can be built over time. Licensing on a company level is then guided by the agreement, serving as a framework enabling companies to know where policy is heading, also on the longer term."

All respondents thus got a similar stimulus clarifying what the basic idea of negotiated agreements is all about.

In the next section we present the data from this survey in connection with the data from the evaluation study on the Dutch negotiated agreements (this is the study on environmental negotiated agreements, for a study on energy efficiency negotiated agreements, see Bressers, de Bruijn, & Dinica, 2007). This way the attitudes and expectations of the Mexican business leaders can be compared with the real results obtained from Dutch practice.

14.5 Results and Discussion of Data

14.5.1 Evaluating Mexican Environmental Policy and Options for Improvement

When asked the question "What is your opinion about the Mexican environmental policy instruments for improving the environmental performance of companies?" no one answered that the existing environmental policy instruments are sufficient and reasonably well implemented. That "the existing environmental policy instruments are as such sufficient, but not consistently enough implemented to get an equal competition situation", was adhered to by 11 respondents, while 5 even think that "the existing environmental policy instruments are not sufficient or are inapt to be implemented well in the Mexican situation". All of these five are members of company committees. The two that are members of government advisory boards both hold the last opinion.

14.5.1.1 Better Instrumentation

We asked them to choose one or more options (with allowing them to add more options themselves) to improve implementation or instrumentation. Of the five that think instrumentation is lacking all want additional economic instruments, three want more or different regulative instruments, two more information instruments and three (plus one that sees implementation is lacking) more negotiated agreements. One respondent added the importance of social responsibility. Another respondent remarks that: "The 'agreements' are an excellent option. The majority of the productive sectors should be included and especially the small and medium industry should be aligned to the 'agreements' since this is the type of industry that normally is not considered in the projects of environmental politics and also produces more damage or do not accomplish the norms. On the other hand there should be other economic instruments that help that the projects to be viable."

14.5.1.2 Improving Implementation

Thirteen people filled in the questions on how to improve implementation, including two of the five that blamed instrumentation in the first place. Of these respondents all but one saw an important role for negotiated agreements as a framework for guiding implementation. Using a combination of grants and requirements of local authorities was also mentioned often, which coincidentally was the way the efforts to improve implementation started in the Netherlands before the negotiated agreements and continued all through the 90s (Bressers, 2004). One respondent remarks: "There should be a higher diffusion of support programs from the federal government and demanding of development for Environmental Politics programs to municipal and state level; continuity of support programs, it means that those programs remain in despite of change of governors." The issue of political support is carried particularly by those who have worked for a relatively short time in this field for their companies.[1] Not surprisingly the need for more political support also often coincides with the desire for more local capacity building.[2]

Here follows an overview of the questions and answers given:
"If you consider only the present implementation of policy instruments lacking, please indicate what policy changes could improve this situation (you may tick more than one)":

– 8x More grants for local authorities to hire good staff, combined with obligatory reporting on implementation to higher authorities

---

[1] Spearman's Rho is 0.727, $p=0.002$, $n=13$. Spearman's Rho is a correlation coefficient for data on ordinal level. With these small numbers each correlation was also checked with cross tabulations. P is the statistical significance (the likelihood of the relation being just a coincidence). N is the number.

[2] Rho 0.415, $p=0.079$, $n=13$

- 2x More obligatory public transparency of business concerning resource use and emissions
- 6x Clearer political support from higher authorities to take environmental law seriously
- 12x Creating a negotiated agreement per state and/or sector of industry that specifies priorities and creates an agreed framework for implementation
- 3x Other, please describe ...

The "other" ideas were often connected to the functioning of the political system. One wrote: "More coordination among the different levels of government (authorities)". Another: "Better distribution of responsibilities and attributions in the government across the different levels (federal, state and municipal). Also, a higher efficiency in the transversal coordination (inter ministries); creation of visible environmental incentives for the best environmental performance." A third one: "Elimination of corruption within the authorities" and "The application of the Environmental Law to all size of industries (large, medium and small). In the current situation only the large and some medium industries are inspected."

All in all, 14 of 16 favour the use of negotiated agreements for one of those two purposes, mostly to support implementation and not as a "stand alone" instrument. This is very interesting, since most Dutch examples of environmental negotiated agreements are also not stand alone instruments.

14.5.2 Expectations to Be Met by Negotiated Agreements Ambition of Negotiated Agreements

One could wonder whether the respondents see the negotiated agreements approach as only a "soft" and business friendly way of environmental policy; a way to avoid and postpone real environmental improvements. Therefore we asked them to express their views on how serious the agreements should be (Table 14.4).

Table 14.4 Responses to "When government would conclude a multi-year negotiated agreement with your sector of industry what would be necessary objectives to make this worthwhile?"

	Entirely Agree	Agree	Neutral	Disagree	Entirely disagree
Ambition: the objectives should be beyond existing regulation	10	4	1		1
Ambition: the objectives should be beyond "business as usual"	7	8			1
Ambition: the objectives should imply real ecological innovation	4	8	3		1
Compliance: the agreements should be guarded against "free riders" that spoil the joint effort	10	5		1	

Only one of the respondents clearly wanted to avoid high ambitions, except in the case of guarding against free-riders. This last subject was deemed unimportant by one other respondent. All others hold the opinion that negotiated agreements only make sense when they really further environmental improvements. The interviewees with a more technical than administrative function seem to be more restrictive (or pessimistic) on striving for real ecological innovations.[3] The respondents see the existing Mexican regulation as the weakest ambition, even weaker than "business-as-usual". This is unlike the Netherlands' study, where the ambition of regulation was seen as beyond business-as-usual.

14.5.2.1 Efficiency of Negotiated Agreements

In addition to the environmental results, the efficiency of the effort is a central goal of the negotiated agreement approach. We also asked how important several efficiency aspects are in their viewpoints (Table 14.5).

Table 14.5 Responses to "When government would conclude a multi-year negotiated agreement with your sector of industry what would be the necessary efficiency gains to make this worthwhile?"

	Entirely agree	Agree	Neutral	Disagree	Entirely disagree
General efficiency: minimisation of total costs	7	6	2	1	
1. Better allocation of efforts among companies to lower costs	6	5	5		
2. Better phasing of objectives and measures in time	4	10	1	1	
3. Decrease bureaucratic and administrative costs	7	5	2	2	
4. Support development of new methods and technologies	10	4	1	1	

Of course it is not surprising that the efficiency gains of the negotiated agreements are generally seen as important. It is interesting that this is clearly less outspoken when redistribution of efforts among companies is involved or when decreasing administrative costs is considered. Strongest is the hope that new methods and technologies will be supported, paving the way for *ecological innovation*.

The need for increased overall efficiency has the strongest support among the technically oriented respondents.[4] Respondents that work longer than 6 years in the company are somewhat more relaxed than the others in assessing the need for efficiency in general, and in administrative efficiency in particular.[5] Administrative

---

[3] Rho 0.378, $p=0.073$, $n=16$
[4] Rho 0.727, $p=0.001$, $n=16$
[5] Rho 0.533, $p=0.017$, $n=16$

efficiency is also stressed by people that see more political support as a solution for implementation problems.[6] There is a relation between the evaluation of Mexican environmental policy and the assessment of the need that negotiated agreements should contribute to efficiency.[7] The respondents that see more fundamental problems with the policy than implementation problems alone think less strongly about efficiency gains. By and large one could claim that the Dutch performance as presented in Section 14.2 makes the desired efficiency gains appear to be realistic goals. This is however somewhat less true than was the case with the ambition of the agreements.

14.5.2.2 Positive Side-Effects of Negotiated Agreements

The Dutch study also revealed the large impact of the use of negotiated agreements on "the policy resource base" (De Clercq, 2002). Numerous positive side effects can contribute to the feasibility of further future steps. To what degree do the surveyed Mexican business leaders deem those important?(Table 14.6)

Again it is not surprising that large majorities find the list of positive side effects worthwhile. Issues that regard the coherence of policies seem to have the largest

Table 14.6 Responses to "When government would conclude a multi-year negotiated agreement with your sector of industry what would be desirable side effects to make it extra worthwhile?"

	Entirely Agree	Agree	Neutral	Disagree	Agreeing in Dutch study (%)
Improved target group attitude on the environment	8	7	1		74
More mutual understanding between partners	4	9	3		78
Improved collaboration between government and business	9	7			80
More knowledge on options for environmental improvements	8	7	1		69
Contributions to future env. policy development	8	8			64
Product or process innovations	8	5	2	1	55
New methods & technologies	8	8			44
More coherence in environmental policies regarding industry	10	5	1		77
More harmonisation between environmental and other policies regarding industry	11	4		1	64

[6] Rho 0.570, $p=0.021$, $n=13$

[7] Rho 0.739, $p=0.001$, $n=16$

support: among all industry policies, among environmental policies and in general the cooperation between government and business. When correlated with the characteristics of the business leaders there seems to be some relation between experience (the length of the employment) and the importance attached to more mutual understanding.[8]

That these kinds of expectations from a negotiated agreement approach could be realistic is shown by the results of the Dutch study. In the table the total of 'entirely agree' and 'agree' (in percentages) is listed in the last column (the 'entirely disagree' column was empty). From the Dutch results it is clear that not only the "new product and process innovations" (which were the least strongly desired among Mexican business leaders), but also the "new methods and technologies" show relatively weak performances. In all other cases Dutch practice was however quite encouraging towards the wishes of the Mexican business leaders.

14.5.3 Feasibility of Negotiated Agreements in Mexico

We compared Mexican desires surrounding the negotiated agreement approach with Dutch practice. But how realistic is this? Their feasibility and success are not only a matter of support among business leaders, but also a matter of favourable conditions. In both the European Neapol study and the Dutch evaluation four explanatory factors for negotiated agreement success were theoretically derived and empirically assessed. This analysis supported the value of these factors to explain the success of negotiated agreements (Bressers & de Bruijn, 2005a). These factors are listed in the table below. In their analysis of the feasibility of negotiated agreements in China, Bressers and Xue (2007) also included two additional factors in the wider economical and cultural contexts, something that we cannot repeat here in the setting of this chapter (Table 14.7).

The respondents are quite optimistic about the situation regarding the four conditions that are important for the success of negotiated agreements as studied and confirmed in previous studies. This is the least true for the willingness of the authorities to exert pressure by new alternative instruments when cooperation fails. Here the respondents obviously doubt whether this would be so in their case. For all, their public image is regarded as economically important and worth protecting. Almost all see their sector as well enough represented to be able to negotiate. Even the issue of a basic level of trust between government and industry – that some doubt to be part of the Mexican societal and political culture – is regarded by the respondents as quite favourable in the cases of their industries. This is also related to the experience of the respondents. The longer the respondent has worked for the company the more favourable the level of trust between industry and government in their sector is assessed.[9] People that are more positive about the level of trust are significantly less

---

[8] Rho 0.317, $p=0.115$, $n=16$
[9] Rho 0.436, $p=0.036$, $n=16$

Table 14.7 Responses to "To what degree do you think that the following favourable conditions for the successful application of negotiated agreements are met in the case of your sector of industry?"

	Entirely agree	Agree	Neutral	Disagree	Entirely disagree
There is already a basic level of respect and trust in "fair play" between the sector and government	5	7	3	1	
The sector is homogeneous or has a small number of companies or has a sector organisation that is well respected by the companies	5	8	2	1	
The sector is directly or indirectly producing for consumers and thus concerned about its public image	8	8			
The authorities seem to be prepared to use other instruments than negotiation to get the sector improving its environmental performance	4	3	5	3	1

inclined to see more political support as the solution to implementation problems.[10] People that assess the preparedness of government to use alternative instruments as a threat if necessary attach more importance to improving mutual cooperation as a side effect of negotiated agreements.[11]

The figures that were presented in Section 14.2 do not indicate that the Dutch circumstances were dramatically more favourable than in the Mexican as viewed by the respondents in our survey. There is however one exception. It is not the "cultural" factor of trust, but the "political" one, on the preparedness to use a "stick behind the door" in case the negotiations fail that is regarded with some doubt among our Mexican respondents.

14.6 Summary and Conclusions

In this chapter we started by identifying some issues in the literature on collaboration and innovation for sustainability. More specifically we studied the literature on negotiated agreements and some empirical studies that were held in Europe and the Netherlands, which has used negotiated agreements to a large extent for this purpose, more than any other country). Following this we described Mexican environmental management and the present role of voluntary approaches therein. While ISO 14000 remains the most commonly used format for voluntary efforts by companies in Mexico, there is also a national scheme "Industria Limpia", or Clean

---

[10] Rho −0.780, $p=0.001$, $n=13$
[11] Rho 0.464, $p=0.036$, $n=16$

Industry, that is witnessing increased participation, especially among the larger companies. Though this approach is voluntary, participation is by no means non-committal. This shows that there is some openness in Mexico regarding the options for non-regulatory environmental policy.

So, a next step towards the use of negotiated agreements as a collaborative strategy towards sustainability is not unimaginable. Consequently we studied attitudes towards such an approach on the basis of a survey among 16 Mexican business leaders that have responsibilities in the environmental management of their companies. None of them evaluated Mexican environmental policy as sufficient. To improve implementation of negotiated agreements as a framework to guide implementation is widely supported. Expectations regarding ambition, efficiency gains and positive side effects of the respondents look quite realistic given the practical results as assessed in a study on the Dutch experience with negotiated agreements. The feasibility and success of negotiated agreements is in theoretical and empirical literature explained by four factors. The respondents assess these factors as quite favourable in the situation of their sectors of industry in Mexico.

As a practical consequence, experimenting with negotiated agreements in the Mexican context seems to be a viable and interesting possibility. Maybe such an experiment could be seen as a logical extension of the ongoing "Clean Industry" voluntary program: from voluntarism to real collaboration for ecological innovation!

Acknowledgments We thank Alejandro Sosa, head of the Mexican office of IGEMI (Global Environmental Management Initiative), for his support in the survey distribution and the very helpful discussions which aided in the formulation of this chapter. Our gratitude to Raul Tornel, head of the "Subprocuradoria de Auditoria Ambiental" in the Environmental Mexican Ministry for his participation in the GIN symposium (Mexico City, 2007). We thank Cheryl de Boer (CSTM) for the re-edition work.

References

Alvarez-Larrauri, R., & Fogel, I. (2008). Environmental audits as a policy of state: 10 years of experience in Mexico. *Journal of Cleaner Production, 16*, 66–74.

Berchicci, L., & King, A. (2007). Postcards from the edge: A review of the business and environment literature. *The Academy of Management Annals, 1*, 513–547.

Börkey, P., & Lévêque, F. (2000). Voluntary approaches for environmental protection in the European Union – A survey. *European Environment, 10*(1), 35–54.

Bressers, H. T. A. (2004). Understanding the implementation of instruments: How to know what works, where, when and how In M. Lafferty William (Ed.), *Governance for sustainable development: The challenge of adapting form to function.* Cheltenham: Edward Elgar.

Bressers, H. T. A., & de Bruijn T. J. N. M. (2005a). Conditions for the success of negotiated agreements: Partnerships for environmental improvement in the Netherlands. *Business Strategy and the Environment, 14*, 241–254.

Bressers, H. T. A., & de Bruijn T. J. N. M. (2005b). Environmental voluntary agreements in the Dutch context. In C. Edoardo (Ed.), *The handbook of environmental voluntary agreements: Design, implementation and evaluation issues* (pp. 261–281), Dordrecht: Springer.

Bressers, J. T. A., de Bruijn, T. J. N. M., & Dinica, V. (2007). Integration and communication as central issues in Dutch negotiated agreements on industrial energy efficiency. *European Environment, 17*, 217–230.

Bressers, H. T. A., de Bruijn, T. J. N. M., & Lulofs, K. (2009). Evaluation of environmental negotiated agreements in the Netherlands. *Environmental Politics, 18*(1), 58–77.

Bressers, J. T. A., & Xue, Y. (2007). The feasibility of environmental negotiated agreements in China. *International Journal Environment and Sustainable Development, 6* (3), 221–241.

Carraro, C., & Lévêque, F. (Eds.). (1999). *Voluntary approaches in environmental policy.* Dordrecht: Kluwer Academic.

Croci, E. (Ed.). (2005). *The handbook of environmental voluntary agreements: Design, implementation and evaluation issues.* Dordrecht: Springer.

Dasgupta, S., Hettige, H., & Wheeler, D. (2000). What improves environmental compliance? Evidence from Mexican industry. *Journal of Environmental Economics and Management, 39*(1), 39–66.

De Bruijn, T. (2003). Transforming regulatory systems: Multilevel governance in a European context. PIK Report (80), 288–295.

De Bruijn, T. J. N. M., & Norberg-Bohm, V. (2005) *Industrial transformation; voluntary, collaborative and information-based approaches in environmental policies in the US and Europe.* Cambridge, MA: MIT Press.

De Clercq, M. (Eds.). (2002). *Negotiating environmental agreements in Europe: Critical factors for success.* Cheltenham: Edward Elgar.

DeLeon, P., & Riviera, J. (Eds.). (2007). Voluntary environmental programs: A symposium. *Policy Studies Journal, 35* (4), 685–792.

DeLeon, P., & Riviera, J. (2008). Voluntary environmental programs: Are carrots without sticks enough? *The Policy Studies Journal, 36*(1), 61–63.

Delmas, M. A., & Terlaak, A. K. (2001). A framework for analyzing environmental voluntary agreements. *California Management Review, 43* (3), 44–63.

EEA. (1997). *Environmental agreements environmental effectiveness* (Environmental Issues Series No. 3 vol. 1). Copenhagen.

Federal Environmental Protection Office (PROFEPA). (2008). Web site: www.profepa.gob.mx.

Federal and General Environmental Law from the Mexican Ministry of the Environment.(2008) http://www.semarnat.gob.mx/leyesynormas/Pages/leyesdelsectorfederal.aspx.

Glasbergen, P. (1998). Partnership as a learning process. In P. Glasbergen (Ed.), *Co-operative environmental governance* (pp. 133–156). Dordrecht: Kluwer.

Jordan, L., Wurzel, R., & Zito, A. (2003). New instruments of environmental governance. *Environmental Politics, 12*(1), 1–224 (special issue).

Klok, P. J. (1989). Convenanten als instrument van milieubeleid: de totstandkoming en effectiviteit van acht produktgerichte milieuconvenanten en hierop gebaseerde verwachtingen omtrent de effectiviteit van convenanten. Enschede: Universiteit Twente, Faculteit der Bestuurskunde. ISBN: 9036502942.

Lyon, T. P., & Maxwell, J. W. (2003). Self-regulation, taxation and public voluntary environmental agreements. *Journal of Public Economics, 87*, 1453–1486.

Medina-Ross, V. (2003). *Institutional and organizational context for corporate voluntary environmental initiatives (VEI) in Mexico.* 11th Conference of GIN, San Francisco.

Mol, A., Lauber, V., & Liefferink, D. (Eds.). (2000). *The voluntary approach to environmental policy: Joint environmental policy-making in Europe.* London: Oxford University Press.

Mumme, S. (1998). *Environmental politics and policy in Mexico. Ecological policy and politics in developing countries.* New York: State University of New York Press.

OECD. (1998). *Environmental performance review of Mexico.* OECD.

OECD. (2003). *Voluntary approaches for environmental policy – Effectiveness, efficiency and usage in policy mixes.* Paris: OECD.

Orts, E. W., & Deketelaere, K. (Eds.). (2001). *Environmental contracts: Comparative approaches to regulatory innovation in the United States and Europe.* The Hague: Kluwer Law International.

Porter, M., & Van der Linde, M. (1995). Green and competitive: Ending the Stalemate. *Harvard Business Review, September–October*, 120–134.

Press, D. (2007). Industry, environmental policy, and environmental outcomes. *Annual Review of Environment and Resources*, *32*, 317–344.

PROFEPA. (2008). Programa Nacional de Auditoria Ambiental. http://www.profepa.gob.mx/PROFEPA/AuditoriaAmbiental/.

Rennings, K., Brockman, K., & Bergmann, H. (1997). Voluntary agreements in environmental protection: Experiences in Germany and future perspectives. *Business Strategy and the Environment*, *6*, 245–263.

Sarkar, R. (2008). Public policy and corporate environmental behavior: A broader view. *Corporate Social Responsibility and Environmental Management*, *15*, 281–297.

Ten Brink, P. (Ed.). (2002). *Voluntary environmental agreements: Process, practice and future use*. Sheffield: Greenleaf.

Tornel, R. (2007). SUBPROCURADURÍA DE AUDITORÍA AMBIENTAL, presentation at GIN Conference Simposio Internacional "El Agua y el Cambio Climático" Mexico, ITESM-CEM-CSTM-GIN.

Index

Note: Locators followed by 'f' and 't' refer to figures and tables respectively.